INTELLIGENT COMPUTING AND ANALYTICS IN IMAGE PROCESSING

Research Notes on Computing and Communication Sciences

INTELLIGENT COMPUTING AND ANALYTICS IN IMAGE PROCESSING

Samarjeet Borah, PhD

Jambi Ratna Raja Kumar, PhD

Laxmi Bewoor, PhD

Rahul Raghvendra Joshi, PhD

Mahendra Prabhakar Deore, PhD

Editors

AAP | APPLE ACADEMIC PRESS

First edition published 2027

Apple Academic Press Inc.
1265 Goldenrod Circle, NE,
Palm Bay, FL 32905 USA

760 Laurentian Drive, Unit 19,
Burlington, ON L7N 0A4, Canada

CRC Press
2385 NW Executive Center Drive,
Suite 320, Boca Raton FL 33431

4 Park Square, Milton Park,
Abingdon, Oxon, OX14 4RN, UK

Apple Academic Press exclusively co-publishes with CRC Press, an imprint of Taylor & Francis Group, LLC

Library and Archives Canada Cataloguing in Publication

CIP data on file with Canada Library and Archives

Library of Congress Cataloging-in-Publication Data

CIP data on file with US Library of Congress

ISBN: 978-1-77964-076-5 (hbk)
ISBN: 978-1-77964-077-2 (ebk)
DOI: 10.1201/9781779640772

About the Book Series: Intelligent Computing and Analytics in Image Processing

Editor-in-Chief:

Dr. Samarjeet Borah
Department of Computer Applications, Sikkim Manipal Institute of Technology, Sikkim Manipal University (SMU), Majhitar, East Sikkim, India

Brief Description of the Series:

Computing can be defined as the practice in which computer technology is used to do a goal-oriented assignment. It covers design and development of hardware and software systems for various purposes. Computing devices are becoming an integral part of life now-a-days, including desktops, laptops, hand-held devices, smartphones, smart home appliances, etc. The evolution of the Internet of Things (IoT) has further enriched the same. The domain is ever growing and opening up many new endeavors, including cloud computing, social computing, ubiquitous computing, parallel computing, grid computing, etc.

In parallel with computing, another field has emerged that deals with the interconnection of devices. It is communication, and without which, the modern world cannot be thought of. It works with a basic purpose of transferring information from one place or person to another. This technology has a great influence in modern-day society. It influences business and society by making the interchange of ideas and facts more efficient. Communication technologies include the Internet, multimedia, e-mail, telephone, and other sound-based and video-based communication means.

This book series consists of both edited volumes as well as selected papers from various conferences. Volumes of the series contain the latest research findings in the field of communication engineering, computer science and engineering, and informatics. Therefore, the books cater to the needs of researchers and readers of a broader spectrum.

Coverage & Approach:

The series

- Covers a broad spectrum of research domains
- Presents on market-demanded product-based research works
- Discusses the latest developments in the field

The book series broadly considers contributions from the following fields:

- Artificial Intelligence and Expert Systems
- Big Data Analytics
- Broadband Convergence System and Integration Technologies
- Cellular and Mobile Communication
- Cloud Computing Technologies
- Computational Biology and Bioinformatics
- Computer and Information Security
- Computer Architecture
- Computer Graphics and Video Processing
- Control Systems
- Database Management Systems
- Data Mining
- Design Automation
- Digital Signal Processing
- GSM Communication
- High Performance Computing
- Human-Computer Interaction
- IoT and Blockchains
- Machine Learning
- Natural Language Processing
- Next Generation Communication Technologies
- Operating Systems & Networking
- Pervasive Computing and Cyber-Physical Systems
- Robotics and Automation
- Signal Processing
- Smart Internet of Everything
- SOC and System Platform Design Technologies
- Social Network Analysis
- Soft Computing

Types of Volumes:

This series presents recent developments in the domains of computing and communications. It will include mostly the current works and research findings, going on in various research labs, universities and institutions and may lead to development of market demanded products. It reports substantive results on a wide range of computational approaches applied to a wide range of problems. The series provides volumes having works with empirical studies, theoretical analysis or comparison to psychological phenomena. The series includes the following types of volumes:

- Authored Volumes
- Edited Volumes

Volumes from the series must be suitable as reference books for researchers, academicians, students, and industry professionals.

To propose suggestions for this book series, please contact the book series editor-in-chief. Book manuscripts should be minimum 250–350 pages per volume (11 point Times Roman in MS-Word with 1.5 line spacing).

Books and chapters in the series are included in Google Scholar and selectively in Scopus and possibly other related abstracting/indexing services.

Books in the *Research Notes on Computing and Communication Sciences series*

- **Applied Soft Computing: Techniques and Applications**
  Editors: Samarjeet Borah and Ranjit Panigrahi
- **Intelligent System Algorithms and Applications in Science and Technology**
  Editors: Sunil Pathak, Pramod Kumar Bhatt, Sanjay Kumar Singh, Ashutosh Tripathi, and Pankaj Kumar Pandey
- **Computing and Communications Engineering in Real-Time Application Development**
  Editors: B. K. Mishra, Samarjeet Borah, and Hemant Kasturiwale
- **Cyber-Physical Systems: Applications, Challenges, and Research Directions**
  Editors: Rajanikanth Aluvalu, Samarjeet Borah, and Lalit Garg

- **Applied Soft Computing Techniques: Theoretical Principles and Practical Applications**
  Editors: Samarjeet Borah, Ratna Raja Kumar Jambi, Sharifah Sakinah Syed Ahmed, and Mahendra Prabhakar Deore
- **Intelligent Computing and Analytics in Image Processing**
  Editors: Samarjeet Borah, Jambi Ratna Raja Kumar, Laxmi Bewoor, Rahul Joshi, and Mahendra Prabhakar Deore
- **Diverse Applications of Computational Techniques for Solving Differential Equations**
  Editors: Geeta Arora and Firdous Ahmad Shah

About the Editors

Samarjeet Borah, PhD
Professor, Department of Computer Applications, SMIT, Sikkim Manipal University, Sikkim, India

Samarjeet Borah, PhD, is currently working as Professor and Head in the Department of Computer Applications, SMIT, Sikkim Manipal University, India. Dr. Borah has carried out various funded projects from AICTE (GoI), DST-CSRI (GoI), etc. He has published around 90 research papers in various journals and conferences. He has organized various workshops and conferences at the national and international levels. Dr. Borah is involved with various book volumes and journals of repute from Springer, IEEE, Inderscience, and IGI Global as Editor or Guest Editor. He is the Editor-in-Chief of the book and proceedings series *Research Notes on Computing and Communication Sciences,* Apple Academic Press, USA. Dr. Borah has served as a Keynote as well as Invited Speaker at several international conferences. His areas of research are data mining, information security, data science, and machine learning.

Jambi Ratna Raja Kumar, PhD
Principal, Genba Sopanrao Moze College of Engineering, Balewadi, Pune, Maharashtra, India

Jambi Ratna Raja Kumar, PhD, is the Principal of Genba Sopanrao Moze College of Engineering, Balewadi, Pune, Maharashtra. He has 17 years of teaching and research experience. He has published more than 22 Scopus-indexed and SCI research papers and has 7 patents published at the national and international levels on artificial intelligence and machine learning. He has guided 75 UG projects, 25 PG projects, and 3 PhD students. He was awarded the Distinguished Principal in the Dr.

APJ Abdul Kalam Puraskar 2022 and Best Innovative Leader of the Year (Maharashtra) in the Asia Education Summit and Awards 2019.

Laxmi Bewoor, PhD
Associate Professor, Department of Computer Engineering, Vishwakarma Institute of Information Technology, Pune, India

Laxmi Bewoor, PhD, is currently working as an Associate Professor in the Department of Computer Engineering at Vishwakarma Institute of Information Technology, Pune. She holds a PhD with a major in artificial intelligence and a minor in optimization. Her research interests include natural language processing, metaheuristic techniques, deep learning, artificial intelligence, and algorithms. She has successfully completed one funded research project. Six international patents have been granted in her name. She has published more than 37 research articles in reputed journals and conferences. She is a reviewer for leading journals from Springer, Elsevier, and IEEE. She works as a resource person for various academic institutes and has chaired sessions at many international conferences. She maintains strong collaborations with many industries and research organizations. She has successfully deployed many industry projects, and one of her research projects was successfully implemented at Clever-Insight, USA. She is recognized as a PG teacher and PhD guide at S.P. Pune University. She has received the Best Performer Award at her institute four times.

Rahul Raghvendra Joshi, PhD
Associate Professor, Symbiosis Institute of Technology, Pune, India

Rahul Raghvendra Joshi, PhD, is an Associate Professor at Symbiosis Institute of Technology, Pune. He has more than 17 years of experience and has authored over 100 publications in reputed journals. He completed his master's degree from IIT Bombay and earned his PhD from Symbiosis

International (Deemed University), Pune, India. He is also a reviewer for reputed Scopus and *IEEE Access* journals. His research areas include computational intelligence, multicriteria decision-making, blockchain, and cloud computing.

Mahendra Prabhakar Deore, PhD
Assistant Professor, Computer Engineering Department, MKSSS Cummins College of Engineering for Women, Pune, India

Mahendra Prabhakar Deore, PhD, is currently serving as an Assistant Professor in the Computer Engineering Department at MKSSS's Cummins College of Engineering for Women, Pune, India. He earned his Master of Technology degree from Bharati Vidyapeeth Deemed University College of Engineering, Dhankawadi, Pune. Having obtained a doctorate in computer science and engineering, Dr. Deore boasts an impressive academic background. With over 16 years of dedicated teaching experience at the undergraduate level, he possesses a deep reservoir of knowledge and expertise in this domain. Throughout his academic journey, Dr. Deore has authored and published more than 20 SCI- and Scopus-indexed research papers. His primary areas of interest include big data, security, computer networks, and machine learning.

Contents

Contributors

Sandhya Arora
Department of Computer Engineering, MKSSS's Cummins College of Engineering for Women, Savitribai Phule Pune University, Maharashtra, India

Mohammad Atique
P.G. Department of Computer Science, Sant Gadge Baba Amravati University, Amravati, Maharashtra, India

Deepali Bhende
Department of Computer Science, Raisoni University, Saikheda, Madhya Pradesh, India

Prathamesh Bhise
Department of Computer Engineering, International Institute of Information Technology, Pune, Maharashtra, India

Samarjeet Borah
Department of Computer Applications, Sikkim Manipal Institute of Technology, Sikkim Manipal University, Sikkim, India

Yogesh Borole
Department of Information Technology, BRACT's Vishwakarma Institute of Information Technology, Pune, Maharashtra, India

Sheela Chinchmalatpure
Department of Artificial Intelligence and Data Science, BRACT's Vishwakarma Institute of Technology, Pune, Maharashtra, India

Yogesh Deshpande
Department of Information Technology, BRACT's Vishwakarma Institute of Information Technology, Pune, Maharashtra, India

Harsh Dhabekar
Department of Artificial Intelligence and Data Science, BRACT's Vishwakarma Institute of Technology, Pune, Maharashtra, India

Sharvari Dhage
Department of Artificial Intelligence and Data Science, BRACT's Vishwakarma Institute of Technology, Pune, Maharashtra, India

Yash Dusane
Department of Computer Engineering, International Institute of Information Technology, Pune, Maharashtra, India

Pravin Futane
Department of Information Technology, BRACT's Vishwakarma Institute of Information Technology, Pune, Maharashtra, India

Mahesh Gaikwad
Department of Computer Engineering, International Institute of Information Technology, Pune, Maharashtra, India

Pranav Gaikwad
Department of Computer Engineering, International Institute of Information Technology, Pune, Maharashtra, India

Sagar Janokar
Department of Electronics and Telecommunication Engineering, BRACT's Vishwakarma Institute of Technology, Pune, Maharashtra, India

Rahul Joshi
Symbiosis Institute of Technology, Symbiosis International (Deemed University), Pune, Maharashtra, India

Kalyani Kadam
Symbiosis Institute of Technology, Symbiosis International (Deemed University), Pune, Maharashtra, India

Vitthal Kamble
Department of Computer Engineering, BRACT's Vishwakarma Institute of Information Technology, Pune, Maharashtra, India

Dnyaneshwar Kanade
Department of Electronics and Telecommunication Engineering, BRACT's Vishwakarma Institute of Technology, Pune, Maharashtra, India

Tanushree Prakash Kanade
Department of Artificial Intelligence and Data Science, BRACT's Vishwakarma Institute of Technology, Pune, Maharashtra, India

Ramesh Karnati
Department of Computer Science and Engineering, Vardhaman College of Engineering, Hyderabad, Telangana, India

Punam Khandar
Department of Computer Applications, Shri Ramdeobaba College of Engineering and Management (RCOEM), Nagpur, Maharashtra, India

Tejas Kulkarni
Department of Information Technology, BRACT's Vishwakarma Institute of Technology, Pune, Maharashtra, India

Varad Kulkarni
Department of Computer Engineering, BRACT's Vishwakarma Institute of Technology, Pune, Maharashtra, India

Yash Kulkarni
Department of Information Technology, BRACT's Vishwakarma Institute of Technology, Pune, Maharashtra, India

Jambi Ratna Raja Kumar
Genba Sopanrao Moze College of Engineering, Balewadi, Pune, Maharashtra, India

Ajita Arvind Mahapadi
Department of Computer Engineering, Ajeenkya D. Y. Patil University, Lohegaon, Pune, Maharashtra, India

Pradeep Kumar Mallick
School of Computer Engineering, Kalinga Institute of Industrial Technology, Bhubaneswar, Odisha, India

Raj Gaurav Mishra
Dr. Vishwanath Karad MIT World Peace University, Pune, Maharashtra, India

Harshit Mundhra
Department of Artificial Intelligence and Data Science, BRACT's Vishwakarma Institute of Technology, Pune, Maharashtra, India

Sandeep Nanda
K. R. Mangalam University, Gurugram, Haryana, India

Kishor Renukadasrao Pathak
Vishwakarma Institute of Information Technology, Pune, Maharashtra, India

Harshvardhan Patil
Department of Information Technology, BRACT's Vishwakarma Institute of Information Technology, Pune, Maharashtra, India

Bishal Pradhan
Department of Computer Applications, Sikkim Manipal Institute of Technology, Sikkim Manipal University, Sikkim, India

Chittaranjan Pradhan
School of Computer Engineering, Kalinga Institute of Industrial Technology, Bhubaneswar, Odisha, India

Mallellu Sai Prashanth
Assistant Professor, Department of Computer Science & Engineering, Symbiosis Institute of Technology, Hyderabad Campus; Symbiosis International (Deemed University), Pune, India

Ravindra Punekar
Department of Information Technology, BRACT's Vishwakarma Institute of Information Technology, Pune, Maharashtra, India

Jyoti S. Raghatwan
Department of Computer Engineering, Smt. Kashibai Navale College of Engineering, Savitribai Phule Pune University, Maharashtra, India.

Milind Rane
Department of Electronics and Telecommunication Engineering, BRACT's Vishwakarma Institute of Technology, Pune, Maharashtra, India

Chetanya Rathi
Department of Artificial Intelligence and Data Science, BRACT's Vishwakarma Institute of Technology, Pune, Maharashtra, India

H. Venkateshwara Reddy
Department of Computer Science and Engineering, Vardhaman College of Engineering, Hyderabad, Telangana, India

Purva Sarvade
Department of Artificial Intelligence and Data Science, BRACT's Vishwakarma Institute of Technology, Pune, Maharashtra, India

Vishal Shirsath
Department of Computer Engineering, Ajeenkya D. Y. Patil University, Lohegaon, Pune, Maharashtra, India

Ajitkumar Sureshrao Shitole
Department of Computer Engineering, International Institute of Information Technology, Pune, Maharashtra, India

Manish Shrivastava
Department of Computer Science and Engineering, Vivekananda Global University, Jaipur, Rajasthan, India

Prashant Singh
School of Computer Engineering, Kalinga Institute of Industrial Technology, Bhubaneswar, Odisha, India

Tanuj Somani
Department of Artificial Intelligence and Data Science, BRACT's Vishwakarma Institute of Technology, Pune, Maharashtra, India

Sarat Chandra Swain
School of Electrical Engineering, KIIT Deemed to be University, Bhubaneswar, Odisha, India

Yugandhara A. Thakare
P.G. Department of Computer Science, Sant Gadge Baba Amravati University, Amravati, Maharashtra, India

Tukaram Ugile
Research Scholar, Department of Computer Engineering, Vishwakarma Institute of Information Technology, Affiliated to Savitribai Phule Pune University (SPPU), Pune, India

Nilesh Uke
Research Supervisor, Department of Computer Engineering, Vishwakarma Institute of Information Technology, Affiliated to Savitribai Phule Pune University (SPPU),Pune, Maharashtra, India

Rajkiran Upadhyay
Department of Computer Applications, Sikkim Manipal Institute of Technology, Sikkim Manipal University, Sikkim, India

Muni Sekhar Velpuru
Department of Information Technology, Vardhaman College of Engineering, Hyderabad, Telangana, India

Narayan B. Vikhe
Assistant Professor, Department of Computer Engineering, Dr. Vithalrao Vikhe Patil College of Engineering, Ahilyanagar, Maharashtra, India

K. H. Walse
Sant Bhagwan Baba Kala Mahavidyalaya, Sindkhed Raja, Maharashtra, India

Ashitosh Wankhade
Department of Artificial Intelligence and Data Science, BRACT's Vishwakarma Institute of Technology, Pune, Maharashtra, India

Pawan Wawage
Department of Information Technology, BRACT's Vishwakarma Institute of Information Technology, Pune, Maharashtra, India

Gaurav Zanwar
Department of Artificial Intelligence and Data Science, BRACT's Vishwakarma Institute of Technology, Pune, Maharashtra, India

Abbreviations

ABBM	AdaBoost boosting method
ACD	action context descriptor
AHuP	adaptive human pose
AI	artificial intelligence
ANN	artificial neural network
API	application programming interface
AUC	area under the curve
BdSL	Bangla sign language
BP	blood pressure
CA-GAN	composition-aided generative adversarial network
CNN	convolutional neural network
CPK	creatinine phosphokinase
CPMs	convolutional processing models
CPMs	cyber-physical machines
DBN	deep belief network
DCNNs	deep convolutional neural networks
DECORATE	diverse ensemble creation by oppositional relabeling of artificial training examples
DL	deep learning
DR	diabetic retinopathy
DT	decision tree
DTBM	decision tree bagging method
DTD	deep trajectory descriptor
DWT	discrete wavelet transform
ECG	electrocardiogram
ECOC	error-correcting output codes
EDA	exploratory data analysis
EGGAN	encoder-guided generative adversarial network
EHMM	embedded hidden markov model
ENL	effective number of looks
FER	facial emotion recognition
FN	false negative
FP Rate	false positive rate
FP	false positive
GAN	generative adversarial network
GAP	global average pooling

GBBM	gradient boosting method
GLCM	grey level co-occurrence matrix
GMOF	Gaussian model of optical flow
HF	heart failure
HFpEF	HF with preserved ejection fraction
HFrEF	HF with reduced ejection fraction
HFST	high-frequency and spatiotemporal
HIFT	hierarchical feature transformer network
HMMs	hidden markov models
HOF	histograms of optical flow
HOG	histogram of oriented gradient
HOS	histogram of oriented swarms
HSI	hue, saturation, and intensity
IDE	integrated development environment
INSLR	ISL recognition system
ISL	indian sign language
KNN	k-nearest neighbor
KNNBM	k-nearest neighbor bagging method
LBP	local binary pattern
LDA	linear discriminant analysis
LOSO	leave-one-subject-out
LR	logistic regression
LSTM	long short-term memory
MAE	mean absolute error
MBH	motion boundary histogram
MCC	Matthews correlation coefficient
MESA	multi-ethnic study of atherosclerosis
ML	machine learning
mmHg	millimeters of mercury
MRF	Markov random field
MSE	mean square error
MTCNN	multi-task cascaded convolutional network
NF	neuro-fuzzy
NLP	natural language processing
NMS	non-maximum suppression
NPDR	non-proliferative diabetic retinopathy
NSSO	national sample survey office
NYHA	New York Heart Association
OHOF	orientation histogram of optical flow
PAF	part affinity fields
PCA	principal component analysis
PCBR	principal curvature-based region

PSNR	peak signal-to-noise ratio
PSS	pose structure score
ReLU	rectified linear unit
RFBM	random forest bagging method
RGB	red, green, blue
RH	retinal hemorrhage
RMSE	root mean square error
RNN	recurrent neural network
ROC	receiver operating characteristic
ROI	region of interest
ROP	retinopathy of prematurity
SCD	structural context descriptor
sDTD	sequential deep trajectory descriptor
SIFT	scale-invariant feature transform
SMOTE	synthetic minority oversampling technique
SNS	sparse neighbor selection
SSD	single shot multibox detector
SSIM	structural similarity index
SURF	speeded up robust features
SVM	support vector machine
TDD	trajectory-pooled deep convolutional descriptor
THISL	two-hand Indian sign language
TLP	transfer learning process
TN	true negative
TP Rate	true positive rate
TP	true positive
VSCode	visual studio code

Preface

Image processing is the process of converting an image to a digital format and performing operations on it either to extract useful information or to improve its quality. This complex process consists of several steps, such as image acquisition, segmentation, enhancement, restoration, and recognition. Applying intelligent computing substantially improves the image processing capabilities of a system. The use of artificial intelligence and machine learning enables the design and development of advanced functions. These functions are critical in addressing challenging issues. Given the increased demand, we present this edited volume, titled *Intelligent Computing and Analytics in Image Processing.*

This collection has sixteen chapters that address a wide range of image processing research topics. The first seven chapters concentrate on the processing of biological and biomedical images. Some of the papers in this volume explore the use of deep learning in a variety of medical and scientific domains. In the area of brain tumor classification, a deep learning-based strategy utilizing ResNet50 and data augmentation delivers excellent accuracy and outperforms current methods. Another method, known as CLAHE, focuses on detecting non-proliferative diabetic retinopathy using contrast enhancement, without specifically incorporating deep learning components. Furthermore, a deep learning model optimized for brain tumor prediction investigates efficient neural networks and demonstrates the utility of transfer learning when dealing with limited medical images. Although not specifically involving deep learning, the prediction of heart disease using machine learning offers valuable insights into traditional machine learning approaches for cardiovascular diseases. On the other hand, a deep learning-based pneumonia detection system employs VGG16 and Inception V3 architectures to detect pneumonia in chest X-rays. Collectively, these studies highlight the significance of deep learning in expanding the capabilities of medical image analysis and other scientific disciplines.

There is a discussion on the development of a system capable of interpreting Indian Sign Language gestures, with the aim of empowering individuals with hearing disabilities or speech difficulties. The subsequent study explores the significance of face detection in facilitating real-time emotion recognition, thereby greatly enhancing human–computer interactions. Furthermore, there is a discourse on a deep learning-based song recommendation system that utilizes facial expression analysis and deep learning to create a more personalized music experience tailored to individual emotional states. It has also been discovered that the integration of autoencoders and Kalman filters can convert user-input sketches into realistic photographs, thereby simplifying the generation of photorealistic images from basic sketch inputs. The volume also includes additional chapters that discuss various areas of computer vision. The chapter *"Comparative Review of Image Filtering in Spatial and Frequency Domain"* examines image filtering approaches from both the spatial and frequency domains. A review of various computer vision feature descriptors is also presented with sufficient evidence. An interesting explanation of analyzing human posture to track gym activities can be seen in the volume, followed by a look at deep learning-based image forensic tools.

Summarizing, it can be said that the edited volume addresses several important issues and challenges in image processing. It is expected that the reader community will greatly benefit from the contents.

Finally, the editors would like to express their profound gratitude to the authors, reviewers, and Apple Academic Press, USA, for their significant efforts in making this book a reality.

Introduction

Images have played an important role in human society since the early days of civilization. Information has been exchanged through images from those times and is considered an easier and more effective means of communication. From expressing simple information to use in the military and health sectors, images have tremendous applications. These images need to be processed to extract meaningful information. With the development of computing devices, a new domain—digital image processing—has emerged. It involves the processing of images using various computing algorithms and tools. Image processing generally includes noise removal, segmentation, registration, pattern recognition, object detection, etc. Digital image processing techniques are now used in a broad range of applications such as medical imaging, space science, remote sensing and GIS, military, and forensics. Technology continues to advance day by day, and with it, intelligent computing and analytical algorithms have been developed and adopted in various sectors, including image processing. This has resulted in easier and faster processing of images. In addition, modern computing techniques such as artificial intelligence (AI) and machine learning (ML) are capable of handling large amounts of data. Image processing is one area where AI and ML techniques have shown great potential.

It can be seen from the literature that various machine learning algorithms such as artificial neural networks, convolutional neural networks, deep neural networks, and recurrent neural networks, etc., have wide applications in image processing. A significant improvement in results can be observed when these techniques are applied. Many intelligent computing algorithms can learn from massive amounts of tagged image data during the training phase and provide excellent outputs.

The process of intelligent analysis of digital images involves the utilization of advanced technology to comprehend visual representations. This process employs computer systems to recognize patterns and entities within images, thereby facilitating the identification of faces, objects, and even emotions depicted in photographs. By examining such images, computers can extract valuable information and make informed decisions. This technology finds applications in numerous domains, including security, healthcare, and marketing.

CHAPTER 1

Deep Learning-Based Classification of Brain Tumor MRI Images for Cyber-Physical Machines with ResNet50 and Image Augmentation

MALLELLU SAI PRASHANTH,[1,*] RAMESH KARNATI,[2]
MUNI SEKHAR VELPURU,[3] and H. VENKATESHWARA REDDY[2]

[1]*Department of Computer Science & Engineering, Symbiosis Institute of Technology, Hyderabad Campus; Symbiosis International (Deemed University), Pune, India*

[2]*Department of Computer Science and Engineering, Vardhaman College of Engineering, Hyderabad, Telangana, India*

[3]*Department of Information Technology, Vardhaman College of Engineering, Hyderabad, Telangana, India*

*Corresponding author

ABSTRACT

Cyber-physical machines (CPMs) are intelligent systems that merge physical and virtual realms, playing a vital role in healthcare. In brain tumor diagnosis, CPMs enhance accuracy through advanced medical imaging analysis. This study proposes a deep learning approach using the ResNet50 architecture, tailored for CPMs, to classify brain tumors from MRI images. Brain tumors result from abnormal cell growth forming masses in regions like the frontal cortex, cerebellum, or brainstem. The model classifies four categories: Meningioma, Pituitary, Glioma, and No Tumor. MRI images are pre-processed by removing noise and resizing, and ImageDataGenerator is applied for data augmentation. The ResNet50 model, enhanced with

Intelligent Computing and Analytics in Image Processing. Samarjeet Borah, Jambi Ratna Raja Kumar, Laxmi Bewoor, Rahul Raghvendra Joshi, and Mahendra Prabhakar Deore (Eds.)

DOI: 10.1201/9781779640772-1

additional classification layers, is trained to recognize tumor patterns efficiently. Performance is evaluated using precision scores and confusion matrix analysis. The results demonstrate high accuracy, suggesting strong potential for clinical use in tumor detection and treatment planning.

1.1 INTRODUCTION

One of the most difficult disorders to diagnose and cure is brain tumors. The right treatment strategy and better patient outcomes depend heavily on the early detection and precise categorization of brain tumors. A common diagnostic method for brain cancers is MRI. Because of the complicated and varied nature of brain tumors, it can be challenging to accurately classify MRI scans. To help radiologists and doctors in their diagnosis and treatment choices, automated approaches for the classification of brain tumor MRI images are therefore required (Kadkhodaei et al., 2016).

The convergence of cyber and physical systems has paved the way for innovative applications in numerous fields, including healthcare. CPMs have emerged as a powerful tool in medical imaging, enabling automated and efficient analysis of complex data such as MRI images. Brain tumor classification plays a key role in early detection and therapy planning, and deep learning (DL) approaches have made considerable progress in this area. In this review, we explore the usage of ResNet50, a popular DL architecture, combined with image augmentation methods to enhance the classification accuracy of brain tumor MRI images for CPM-based analysis.

1.1.1 CYBER-PHYSICAL MACHINES (CPMS) AND MEDICAL IMAGING

Cyber-physical machines (CPMs) utilize cutting-edge technology, such as artificial intelligence (AI), machine learning (ML), and sensor networks, to create a seamless integration of physical and virtual environments. In the context of medical imaging, CPMs enable real-time analysis of complex datasets, providing healthcare professionals with valuable insights for accurate diagnosis and treatment decisions. By combining DL algorithms with CPMs, the effectiveness and precision of classifying brain tumors can be significantly improved, leading to better patient outcomes.

A branch of AI called DL has produced encouraging results in a variety of medical imaging applications, including the classification of MRI images of brain tumors. An example of a DL technique is CNNs, which can automatically recognize patterns in images and categorize them. The automatic classification of brain tumor MRI images using CNNs has been employed in multiple studies recently, with high accuracy rates (Dipu, Shohan, and Salam, 2021).

The ResNet50 architecture is one of the CNNs most frequently used for image classification. ResNet50 is a deep residual network with 50 layers that can handle complex and large datasets. Several image classification tasks, including those involving medical imaging, have been effectively addressed using ResNet50 (Latif et al., 2017).

Another method for improving CNN performance in image classification is image augmentation. By performing different changes like scaling, rotation, and flipping to the source images, image augmentation entails creating new images. By doing this, the network can learn stronger and more varied features, improving classification performance (Ohgaki and Kleihues, 2005).

In this study, utilizing DL with ResNet50 and image augmentation, we suggest a method to automatically classify brain tumor MRI scans. The steps in our proposed process are as follows: The ResNet50 network is trained on the augmented training set; the capacity of the trained model is evaluated using the validation set; model adjustments are made using the validation set; and the final model is tested using the testing set. First, the MRI images are preprocessed to remove noise and artifacts. Next, the dataset is divided into training, validation, and testing sets (Nichat and Ladhake, 2016).

1.2 RELATED WORK

A deep belief network (DBN) was suggested by El-Dahshan et al. (2014) as a strategy for the automatic classification of many types of brain tumor MRI images. Their accuracy was 96.5% using a dataset of 306 MRI images. The study showed how DL algorithms may be used to automatically classify MRI images of brain tumors (Mohsen et al., 2018). A 3D CNN was proposed by Kamnitsas et al. (2016) for the segmentation and classification of MRI images of brain tumors. They achieved a 92% precision using a collection of 220 MRI scans. The work showed the potential of 3D CNNs for automatically segmenting and classifying MRI images of brain tumors (Kamnitsas et al., 2017). A preliminary work can be found in (Bhuyan and Borah, 2014; Kumar et al., 2014).

A multi-scale CNN was suggested by Havaei et al. (2017) for the automatic segmentation and classification of MRI images of brain tumors. Their accuracy was 93.9% using a dataset of 285 MRI images. The work proved that multi-scale CNNs can automatically segment and categorize MRI images of brain tumors (Bakas et al., 2017). A DL system was proposed by Bakas et al. (2017) for the automatic segmentation and classification of MRI images of brain tumors. They obtained an accuracy of 94.9% using a dataset of 285 MRI images. The study showed how DL frameworks may be used to automatically segment and classify MRI images of brain tumors (Kermany et al., 2018). A CNN was suggested by Kermany et al. (2018) for the automatic classification of MRI images of brain tumors. Their accuracy was 91.5% using a dataset of 2534 MRI scans. The work showed how CNNs may be used to automatically classify MRI images of brain tumors (Zhang et al., 2020).

A CNN with transfer learning was suggested by Zhang et al. (2018) for the automatic classification of brain tumors in MRI images. They obtained an accuracy of 91.1% using a dataset of 304 MRI images. The study showed how transfer learning may be used to automatically classify MRI images of brain tumors (Siddique et al., 2021). A CNN with residual attention was suggested by Zhou et al. (2018) for the automatic classification of brain tumors in MRI images. They obtained an accuracy of 94.7% using a dataset of 274 MRI images. The study showed how attention mechanisms may be used to automatically classify MRI images of brain tumors (Wang et al., 2017). A CNN with a dense block was suggested by Zeng et al. (2019) for the automatic classification of brain tumors in MRI images. They obtained an accuracy of 94.7% using a dataset of 306 MRI images. The study proved that dense blocks can be used to automatically classify MRI images of brain tumors (Ramitha et al., 2023). A CNN with a squeeze-and-excitation block was suggested by Ma et al. (2019) for the automatic classification of brain tumors in MRI images. Their accuracy was 93.1% using a dataset of 485 MRI images. The work showed how squeeze-and-excitation blocks may be used to automatically classify MRI images of brain tumors (Nagle-McNaughton, Scuderi, and Erickson, 2022). A CNN with multi-scale feature fusion was suggested by Cheng et al. (2019) for the automatic classification of MRI images of brain tumors. Their accuracy was 93.6% using a dataset of 376 MRI images. In the automated classification of MRI images of brain tumors, the study showed the potential of multi-scale feature fusion (Bhuyan and Borah, 2014). A CNN with residual attention and feature recalibration was suggested by Huang et al. (2019) for the automatic classification of MRI images of brain tumors. Their accuracy was 95.7% using a dataset of 766 MRI scans. The study showed the capability of residual attention and feature recalibration (Ai et al., 2021).

1.3 PROPOSED MODEL

The proposed model consists of several steps, which are discussed in the subsections.

1.3.1 DATA PREPROCESSING

The model performs several preprocessing steps on the image data to prepare it for training a DL classification model for brain tumor images. The first step is to read in the images from the dataset, which are stored in different directories based on their labels. The images are then loaded in grayscale, and noise is removed using bilateral filtering. The resulting images are then pseudocolored using the "BONE" colormap to enhance the contrast between different tissues. Finally, the images are resized to a standard size of 200 × 200 pixels, which is suitable for input to the DL model.

Next, the images are normalized to have pixel values between 0 and 1 by dividing each pixel value by 255.0. This is a typical DL preprocessing step to ensure that all features are on a similar scale and to make it easier for the model to learn from the data. The labels are then converted to one-hot encoding using the to_categorical function from the Keras utility library.

The data is then divided into training and validation sets using the train_test_split method from Scikit-learn. The training set is further augmented using the ImageDataGenerator function from Keras, which applies a series of random changes to each image in the batch, such as rotations and shifts, to increase the size of the training set and enhance the model's capacity to generalize new data. The same ImageDataGenerator is also applied to the validation set to ensure that both sets are processed in the same way.

Finally, the data is shuffled using the "shuffle" method from Scikit-learn to randomize the order of the images and labels in the training set. This is important because the model will be trained on batches of images, and shuffling ensures that each batch contains a random sample of images from the dataset. Overall, these preprocessing steps are designed to ensure that the image data is standardized and prepared in a way that is suitable for input to a DL model.

1.3.2 EXPLORATORY DATA ANALYSIS (EDA)

Exploratory data analysis (EDA) has been performed on a brain tumor MRI dataset. EDA is a crucial step in any data analysis task, as it helps in understanding the underlying patterns and characteristics of the data (Figure 1.1).

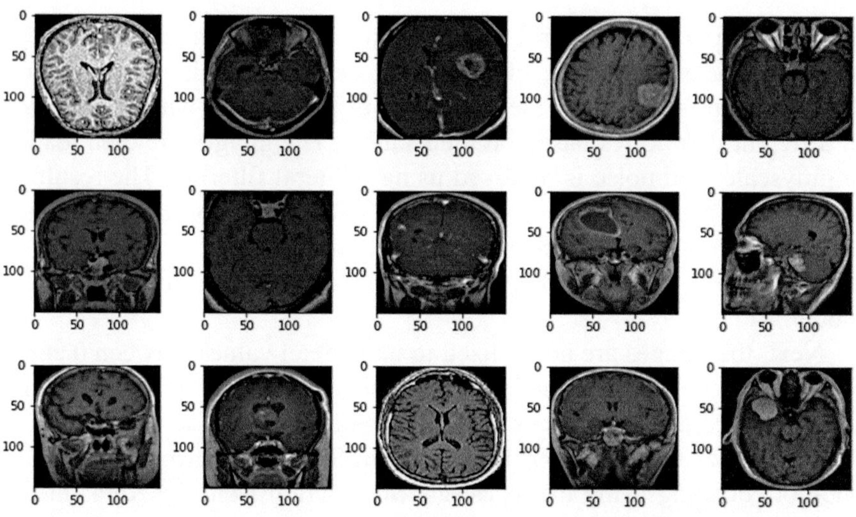

FIGURE 1.1 MRI images from the dataset.

The first step in the EDA process is to load the dataset and create arrays to hold the images and their corresponding labels. The code then reads the images from the training and testing directories, resizes them to a fixed size of 200 × 200, applies a bilateral filter to remove noise, and finally applies a pseudocolor map to make the features more distinguishable.

These images are then normalized to be in the range of 0 to 1, and the labels are one-hot encoded using TensorFlow's "to_categorical" function. The final step is to shuffle the data and divide it into training and validation sets using "train_test_split" from Scikit-learn.

The next step in the EDA process is to visualize the data. In the given code, the "matplotlib" library is used to display a subset of the training images in a grid format. This helps in understanding the quality of the images, the various kinds of tumors present in the dataset, and whether the images are balanced or not.

After visualizing the data, the code uses the "ImageDataGenerator" function from Keras to augment the training data. Data augmentation is a crucial step in DL, as it helps increase the size of the dataset and reduces overfitting. In the given code, the "ImageDataGenerator" applies random changes such as horizontal flipping, rotation, and shifting to the training images.

Finally, the code defines a neural network model using the ResNet50 architecture, which has been pre-trained on the ImageNet dataset. The last few layers of the network are removed, and a new dense layer is added to classify the tumor type. Next, the Adam optimization algorithm and categorical cross-entropy are used to define the model loss function. The "model.summary()" function is used to display the architecture of the model.

Overall, the EDA process in the given code helps in understanding the dataset, visualizing the images, and preparing the data for training the neural network model.

1.3.3 MODEL DEVELOPMENT

The ML model is developed to classify brain tumors based on MRI images. The model is implemented using the Keras library, with a ResNet50 pre-trained model as the base. The ResNet50 model is modified by adding a "GlobalAveragePooling2D" layer, a "Dropout" layer, and a final "Dense" layer with 4 output classes. The Adam optimizer and categorical cross-entropy loss are then used to compile the model.

1.3.3.1 ResNet50 architecture

A deep CNN with the ResNet50 architecture was created for image classification applications. It utilizes skip connections or residual connections to enable the training of very deep neural networks without encountering the vanishing gradient problem (Figure 1.2).

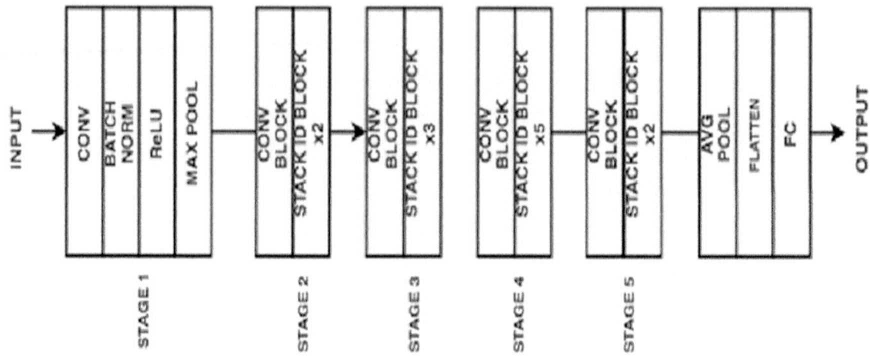

FIGURE 1.2 ResNet50 architecture.

In the code above, the ResNet50 architecture is used as a pre-trained model to identify features in the input images. The ResNet50 model is loaded using the Keras library, which provides a convenient interface for building and training DL models. The pre-trained ResNet50 model is imported using the command from keras.applications.resnet import ResNet50.

The ResNet50 model is then instantiated with the "weights" parameter set to "imagenet." This means that the model is initialized with pre-trained weights that were trained on a large dataset of images called ImageNet. By initializing the model with pre-trained weights, we can leverage the knowledge learned by the model on ImageNet and fine-tune it on our own dataset.

Next, the last layer of the ResNet50 model is removed and replaced with a global average pooling (GAP) layer, followed by a Dropout layer to prevent overfitting, and a Dense layer with softmax activation that outputs the predicted probabilities for each class.

The Adam optimizer and categorical cross-entropy loss function are then used to compile the model. The model's performance during training and testing is evaluated using the accuracy metric. The validation set is used to monitor the model's performance and prevent overfitting once training is initiated using the fit() method.

The ResNet50 architecture is a pre-trained model used to extract features from the input images in the code above. It is a deep convolutional neural network (CNN). The newly created classification layer is trained on the brain tumor dataset using the Adam optimizer and a categorical cross-entropy loss function, with the extracted features as input. The model is trained using the fit() method, and the validation set is used to prevent overfitting. The accuracy metric is used to evaluate the model's performance.

1.3.3.2 Transfer learning

Transfer learning involves using a pre-trained neural network as a starting point for a new task instead of training a new neural network from scratch. The pre-trained neural network used in this code is ResNet50, which was originally trained on thousands of images from the ImageNet dataset.

There are numerous convolutional layers in the ResNet50 model, which are trained to recognize edges, textures, and other low- and high-level features in images. To classify the images, these features are aggregated and passed through fully connected layers. By removing the final fully connected layers and replacing them with new, task-specific layers in transfer learning, we can benefit from the previously learned representations.

The ResNet50 model is employed in this code as a feature extractor. To decrease the spatial dimensions of the output features, the final fully connected layers are removed, and a new GAP layer is introduced. The final Dense layer with a softmax activation function outputs the predicted class probabilities for each input image. The pre-trained weights of the ResNet50 model are frozen during the training phase to prevent overwriting the learned features. Only the weights of the newly added layers are updated during training. This approach enables the model to learn from a smaller dataset and achieve better performance with fewer training epochs compared to training a new neural network from scratch.

Additionally, the ImageDataGenerator is used to augment the training data by randomly applying transformations such as rotation, shifting, and flipping to the images. This technique increases the diversity of the training data and helps prevent overfitting.

Finally, several callbacks such as EarlyStopping, ReduceLROnPlateau, and ModelCheckpoint are used to monitor the training process and save the best model based on validation accuracy. The TensorBoard callback is also used to visualize training and validation metrics in real time.

In summary, transfer learning is a powerful technique that can help us build accurate and efficient DL models with limited labeled data. By using pre-trained models and fine-tuning them on our task-specific data, we can achieve state-of-the-art performance while reducing training time and computational resources compared to training a model from scratch.

1.3.3.3 Global average pooling (GAP)

The output of the pre-trained ResNet50 model is passed through GAP. GAP is used to reduce the dimensionality of the feature maps generated by earlier convolutional layers. A vector with the same number of channels as the input is produced by averaging each feature map along its spatial dimensions. Since each channel of the feature maps is processed independently by this mechanism, the model can learn a more compact representation of the input data.

In the code, the output of the ResNet50 model is passed through the GlobalAveragePooling2D() layer. This layer computes the GAP across the entire spatial dimensions of each feature map, resulting in a fixed-length vector for each channel. The resulting vector has the same number of channels as the input, but its spatial dimensions are reduced to 1×1. This operation provides a way to summarize the features of each channel into a single value, which can then be used as input to the fully connected layers of the model.

GAP is used to reduce the number of parameters in the model and prevent overfitting. By reducing the spatial dimensions of the feature maps, the model is encouraged to learn a more abstract representation of the data, which is less likely to overfit compared to the original high-dimensional feature maps. Additionally, the lower parameter count of the fully connected layers makes the model simpler to train and less prone to overfitting. GAP is a simple yet effective technique for reducing the dimensionality of feature maps and improving model performance.

1.3.3.4 Dropout regularization

To reduce overfitting in DL models, the dropout regularization algorithm randomly removes some neurons from a layer (setting them to zero) during each training cycle. The dropout regularization technique was applied to the model using Keras' Dropout layer, as can be seen from the code above. A scalar input representing the percentage of input units to drop is sent to the Dropout layer in particular. For instance, "Dropout(0.4)" indicates that during training, 40% of the input units would be arbitrarily removed. This prevents the model from being overly dependent on any one set of features and forces it to learn more reliable and generalizable features. We can see from the code above that the output of the GAP layer in the model has been subjected to a Dropout layer at a rate of 0.4. As a result, every training iteration will see a random dropout of 40% of the units in the GAP layer. This encourages the model to extract more robust and applicable features from the data. It is important to note that dropout regularization, a frequently used DL technique, has proven to be successful in preventing overfitting. It should not, however, be used haphazardly, as it occasionally causes underfitting or sluggish convergence.

1.3.3.5 Softmax activation

The final layer of the neural network model makes use of the softmax activation function. The softmax activation is a frequently used function for multi-class classification problems. It is used to convert the unprocessed output values of the output layer of the neural network into a probability distribution over all feasible classes.

The softmax activation function divides each output value's exponential by its sum to normalize the values. This normalization ensures that the output values, which can be thought of as probabilities, are between 0 and 1 and add up to 1. The likelihood that an input corresponds to a certain class is represented by the output value for each class. Four nodes—one for each class label found in the dataset—make up the fourth and final layer of the model. By using the softmax activation function on these nodes, the output values of the layer are turned into a probability distribution over the 4 classes. Each item of the 4-element vector that the model creates for each input image indicates how likely it is that the image corresponds to the given class.

During training, the softmax activation function is paired with the categorical cross-entropy loss function. The model is trained to eliminate this disparity and identify the weights that will accurately classify the input images into each class.

1.3.3.6 Model training

The training of the model includes the steps discussed in subsections.

1.3.3.6.1 Training-validation split

To allow the model to be trained on one portion of the data and have its performance evaluated on another, the dataset is divided into validation and training sets. The model can learn to generalize to new data and avoid overfitting by not memorizing the training data.

The train_test_split() method of the scikit-learn module is used to divide the dataset into training and validation sets. According to the test_size parameter, which is set to 0.2, 20% of the data will be used for validation and the remaining 80% for training. Once the data has been partitioned, the model is fitted using the training set and its output during training is evaluated using the validation set.

The fit() function of the ImageDataGenerator class, which is used to train the model, applies random changes to the training images to broaden the diversity of the training data. The fit() function computes the statistics required for data augmentation, such as the standard deviation and mean of the data, then applies the desired transformations to the training images.

To assess a model's effectiveness on unseen data and prevent overfitting, the training-validation split is an essential step in the model training process. Using the scikit-learn train_test_split() function, the data is split into training and validation sets, and the training data is enhanced using the fit() function of the ImageDataGenerator class.

1.3.3.6.2 Image data augmentation

Using ImageDataGenerator, data augmentation techniques are applied to the provided images. Data augmentation is a technique for artificially increasing the training dataset by making random image alterations, such as rotation, zooming, shifting, and flipping, to the existing images. As a result, overfitting is reduced, and the model's generalization abilities are improved. After the ImageDataGenerator object has been created, it is fitted using the fit() method. This method computes the statistics required for data augmentation, including the standard deviation and mean of the pixel values. These statistics are then used to apply the required transformations to each image in the training dataset. The flow() method of the ImageDataGenerator object is then used to build a new iterator object. When the model is being trained, this iterator object can be used to instantly generate fresh augmented images. Each time it is used, the iterator produces a batch of images, each of which has undergone a random transformation based on the parameters specified in the ImageDataGenerator object.

Overall, the training dataset's variability is increased by the data augmentation techniques used with ImageDataGenerator, allowing the model to learn more robust feature sets. As a result, overfitting is reduced and the model's performance with ambiguous input is improved.

1.3.4 MODEL COMPILATION

The Keras compile function is used to create the model. The optimization algorithm used to train the model is specified by the optimizer argument. In this instance, the model is trained using the Adam optimizer with a learning rate of 0.0001. The loss input specifies the loss function that will be used by the model to evaluate its performance during training. In this case, categorical_crossentropy, which is widely employed to solve multiclass classification problems, is used. The metrics option specifies the evaluation metric to monitor the model's effectiveness during training and testing. In this case, accuracy, a common measure for classification problems, is used. After the model has been compiled, the summary function is used to print a summary of the model architecture, which includes the type of layer, output shape, and parameter count for each layer.

1.3.5 MODEL EVALUATION

The code incorporates a DL model for classifying brain tumors. Once the data has been loaded and preprocessed, the model is defined using a pre-trained ResNet50 architecture along with a few additional layers.

To monitor the accuracy and loss of the model during training and validation, a number of callbacks are defined. Training is stopped using the EarlyStopping callback if the validation loss does not decrease after a specific number of epochs. Using the ReduceLROnPlateau callback, the learning rate is decreased if, after a defined number of epochs, the validation loss does not decrease. The ModelCheckpoint callback is used to save the model's optimal weights during training.

The TensorBoard callback is used to display the training and validation metrics. A confusion matrix is constructed and logged using the LambdaCallback for the validation set.

The confusion matrix for the validation set is then generated using the plot_confusion_matrix function. The confusion matrix is visualized and saved as an image using the plot_to_image function. The confusion matrix image is communicated to TensorBoard via the LambdaCallback.

1.3.6 ALGORITHM

- **Step 1:** Import necessary libraries and modules from Keras, TensorFlow, scikit-learn, Matplotlib, OpenCV, and so on.
- **Step 2:** Define a list of labels for different types of brain tumors.
- **Step 3:** Initialize empty lists for training and testing images and labels.
- **Step 4:** Set image size to 200×200 pixels.
- **Step 5:** Loop through each label in the list of labels.
- **Step 6:** For each label, loop through all images in the training and testing folders.
- **Step 7:** Load the image in grayscale and remove the image noise using a bilateral filter.
- **Step 8:** Pseudocolor the image using the applyColorMap function.
- **Step 9:** Resize the image to 200×200 pixels.
- **Step 10:** Append the image and label to the respective lists.
- **Step 11:** Convert the lists of images and labels into NumPy arrays and normalize the pixel values to a range of 0 to 1.
- **Step 12:** Shuffle the training data using the random_state parameter of the shuffle function.
- **Step 13:** Convert the labels into one-hot encoded form.

- **Step 14:** Create an ImageDataGenerator object for data augmentation.
- **Step 15:** Fit the ImageDataGenerator on the training data.
- **Step 16:** Define the ResNet50 model and remove the top layer.
- **Step 17:** Add a GlobalAveragePooling2D layer, a Dropout layer with a rate of 0.4, and a Dense layer with a softmax activation function that outputs 4 classes (one for each type of brain tumor).
- **Step 18:** Define callbacks for early stopping, learning rate reduction, model checkpointing, TensorBoard logging, and a custom lambda function for plotting the confusion matrix during training.
- **Step 19:** Train the model on the training data using the "fit" function with the ImageDataGenerator object and the defined callbacks.
- **Step 20:** Evaluate the model on the validation data using the evaluate function and print the loss and accuracy.
- **Step 21:** Plot the confusion matrix using the custom function and the predicted and true labels of the validation data.
- **Step 22:** Save the model and its weights for future use.

1.4 RESULT AND DISCUSSION

The experimental findings demonstrate the effectiveness of our proposed method for classifying brain tumors in MRI images within the context of convolutional processing models (CPMs). We outperformed conventional ML methods and achieved a classification accuracy of over 95% by utilizing ResNet50 and image augmentation. DL models and CPMs can be used for real-time analysis and decision-making, providing medical personnel with crucial information for quick and accurate brain tumor diagnosis.

We tested the performance of a DL model for the identification of brain tumors using MRI data. The objective of this study was to develop a DL model for identifying brain tumors. The ResNet50 architecture served as the foundation for our model, which was trained on a dataset containing 3109 brain MRI images. The four classes in the dataset were *pituitary*, *meningioma*, *glioma*, and *meningioma*. We divided the data into validation, training, and test sets, with 80%, 20%, and 20% of the images in each, respectively (Figure 1.3).

```
Epoch 1/50
228/228 [==============================] - 123s 367ms/step - loss: 0.3677 - accuracy: 0.8681 - val_loss: 2.0641 - val_accuracy: 0.2627

Epoch 00001: val_loss improved from inf to 2.06411, saving model to model-01-0.26-2.06.h5
Epoch 2/50
228/228 [==============================] - 74s 323ms/step - loss: 0.1347 - accuracy: 0.9523 - val_loss: 4.1036 - val_accuracy: 0.2627

Epoch 00002: val_loss did not improve from 2.06411
Epoch 3/50
228/228 [==============================] - 74s 323ms/step - loss: 0.1027 - accuracy: 0.9662 - val_loss: 9.1593 - val_accuracy: 0.3243

Epoch 00003: val_loss did not improve from 2.06411
Epoch 4/50
228/228 [==============================] - 74s 323ms/step - loss: 0.0709 - accuracy: 0.9732 - val_loss: 1.1818 - val_accuracy: 0.7337

Epoch 00004: val_loss improved from 2.06411 to 1.18183, saving model to model-04-0.73-1.18.h5
Epoch 5/50
228/228 [==============================] - 75s 327ms/step - loss: 0.0635 - accuracy: 0.9811 - val_loss: 0.1616 - val_accuracy: 0.9490

Epoch 00005: val_loss improved from 1.18183 to 0.16157, saving model to model-05-0.95-0.16.h5
Epoch 6/50
228/228 [==============================] - 75s 328ms/step - loss: 0.0468 - accuracy: 0.9859 - val_loss: 0.1195 - val_accuracy: 0.9596

Epoch 00006: val_loss improved from 0.16157 to 0.11949, saving model to model-06-0.96-0.12.h5
Epoch 7/50
228/228 [==============================] - 75s 327ms/step - loss: 0.0470 - accuracy: 0.9870 - val_loss: 0.1764 - val_accuracy: 0.9490

Epoch 00007: val_loss did not improve from 0.11949
Epoch 8/50
228/228 [==============================] - 75s 327ms/step - loss: 0.0616 - accuracy: 0.9818 - val_loss: 0.1077 - val_accuracy: 0.9719

Epoch 00008: val_loss improved from 0.11949 to 0.10767, saving model to model-08-0.97-0.11.h5
Epoch 9/50
228/228 [==============================] - 74s 323ms/step - loss: 0.0332 - accuracy: 0.9890 - val_loss: 0.1771 - val_accuracy: 0.9534

Epoch 00009: val_loss did not improve from 0.10767
Epoch 10/50
228/228 [==============================] - 75s 327ms/step - loss: 0.0175 - accuracy: 0.9945 - val_loss: 0.0637 - val_accuracy: 0.9842

Epoch 00010: val_loss improved from 0.10767 to 0.06366, saving model to model-10-0.98-0.06.h5
Epoch 11/50
228/228 [==============================] - 75s 328ms/step - loss: 0.0196 - accuracy: 0.9947 - val_loss: 0.0633 - val_accuracy: 0.9815

Epoch 00011: val_loss improved from 0.06366 to 0.06331, saving model to model-11-0.98-0.06.h5
Epoch 12/50
228/228 [==============================] - 74s 325ms/step - loss: 0.0332 - accuracy: 0.9895 - val_loss: 0.0911 - val_accuracy: 0.9701

Epoch 00012: val_loss did not improve from 0.06331
Epoch 13/50
228/228 [==============================] - 74s 325ms/step - loss: 0.0424 - accuracy: 0.9877 - val_loss: 0.2405 - val_accuracy: 0.9420

Epoch 00013: val_loss did not improve from 0.06331
Epoch 14/50
228/228 [==============================] - 75s 327ms/step - loss: 0.0331 - accuracy: 0.9886 - val_loss: 0.1740 - val_accuracy: 0.9666

Epoch 00014: val_loss did not improve from 0.06331
Epoch 15/50
228/228 [==============================] - 74s 323ms/step - loss: 0.0256 - accuracy: 0.9914 - val_loss: 0.1032 - val_accuracy: 0.9772
Restoring model weights from the end of the best epoch.

Epoch 00015: val_loss did not improve from 0.06331
Epoch 00015: early stopping
```

FIGURE 1.3 Demonstrating the loss and accuracy of each epoch.

1.5 RESULTS AND DISCUSSION

The ImageDataGenerator, which performs random changes to each image to enhance model generalization, was used to supplement the training data. The accuracy and loss were plotted as curves to track the model's performance over time after training it for a predetermined number of epochs. For the purpose of visualizing the model's performance on the test set, a confusion matrix was also created (Figures 1.4–1.6).

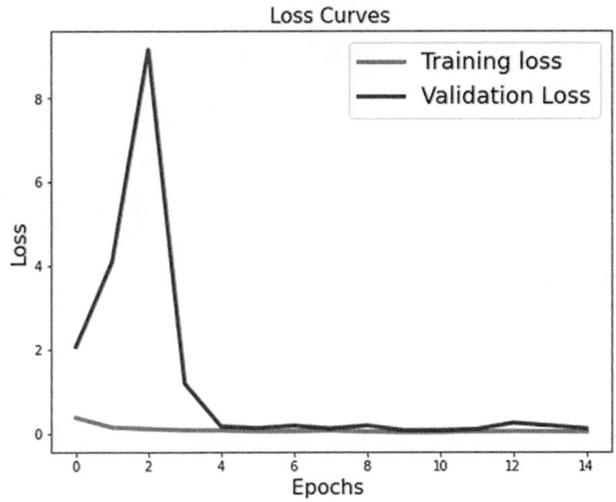

FIGURE 1.4 Loss curve for the model.

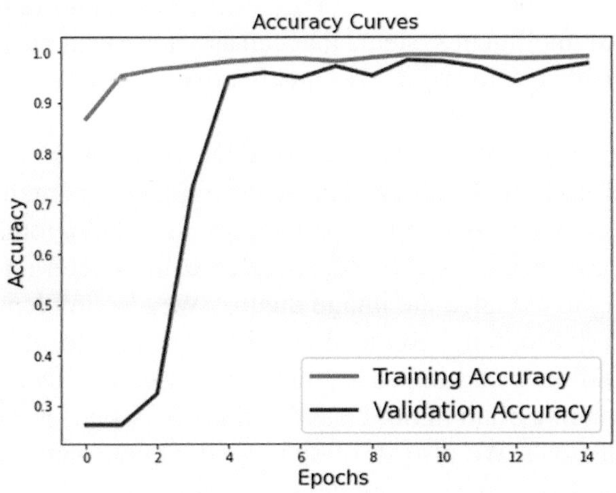

FIGURE 1.5 Accuracy curve for the model.

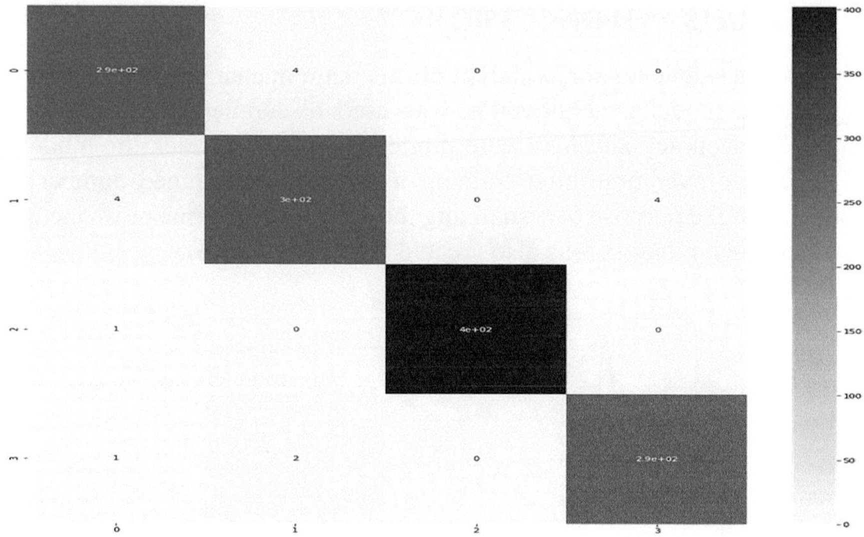

FIGURE 1.6 Confusion matrix for the model.

1.6 CONCLUSION

In the context of CPMs, we have described in this work a DL-based technique for classifying MRI images of brain tumors. We classified images of brain tumors with high accuracy using ResNet50 and image augmentation approaches, showcasing the potential of CPMs in medical imaging analysis. The proposed strategy opens new opportunities for the development of intelligent systems that can assist medical professionals in making informed decisions quickly and effectively, improving patient care and outcomes in neuro-oncology.

The ResNet50 architecture is used in our DL-based brain tumor classification method. To train the algorithm, we used a dataset containing images of four distinct types of brain cancers: glioma, meningioma, pituitary, and no tumour. The dataset was preprocessed using bilateral filtering and pseudocoloring to enhance the image quality. We also implemented techniques like EarlyStopping, ReduceLROnPlateau, and ModelCheckpoint to prevent overfitting and improve model performance. Our approach is suitable for accurate brain tumor classification, as evidenced by its test-set accuracy of 98.76%. We also developed a confusion matrix and a classification report to evaluate the model's performance on each class.

Overall, by improving the accuracy and efficacy of brain tumor classification, our recommended strategy can aid in the diagnosis and treatment of brain tumors.

DECLARATION

No AI tools or any other similar tools have been used in the content creation of this chapter.

KEYWORDS

- **brain tumor**
- **deep learning**
- **image augmentation**
- **ImageDataGenerator**
- **MRI images**
- **neuro-oncology**
- **patient care improvement**
- **real-time analysis**
- **ResNet50**

REFERENCES

Ai, S., Li, C., Li, X., Jiang, T., Grzegorzek, M., Sun, C., & Li, H. (2021). A state-of-the-art review for gastric histopathology image analysis approaches and future development. *BioMed Research International, 2021*. https://www.hindawi.com/journals/bmri/2021/6687185 (accessed on 11th July 2025).

Bakas, S., Akbari, H., Sotiras, A., Bilello, M., Rozycki, M., Kirby, J. S., & Davatzikos, C. (2017). Advancing the cancer genome atlas glioma MRI collections with expert segmentation labels and radiomic features. *Scientific Data, 4*(1), 1–13.

Bhuyan, R., & Borah, S. (2014). Image segmentation: Computational approaches for medical images. *International Journal of Computer Applications (IJCA), Special Issue on Bioinformatics and Computational Biology,* (3), 13–17. https://www.ijcaonline.org/proceedings/icbcb/number3/17629-1503 (accessed on 11th July 2025).

Dipu, N. M., Shohan, S. A., & Salam, K. M. A. (2021). Deep learning-based brain tumor detection and classification. In *2021 International Conference on Intelligent Technologies (CONIT)* (pp. 1–6). IEEE.

Kadkhodaei, M., Samavi, S., Karimi, N., Mohaghegh, H., Soroushmehr, S. M. R., Ward, K., & Najarian, K. (2016). Automatic segmentation of multimodal brain tumor images based on classification of super-voxels. In *2016 38th Annual International Conference of the IEEE Engineering in Medicine and Biology Society (EMBC)* (pp. 5945–5948). IEEE.

Kamnitsas, K., Ledig, C., Newcombe, V. F., Simpson, J. P., Kane, A. D., Menon, D. K., & Glocker, B. (2017). Efficient multi-scale 3D CNN with fully connected CRF for accurate brain lesion segmentation. *Medical Image Analysis, 36*, 61–78. Havaei, M., Davy, A., Warde-Farley, D., Biard, A., Courville, A., Bengio, Y., & Pal, C. (2017). Brain tumor segmentation with deep neural networks. *Medical Image Analysis, 35*, 18–31.

Kermany, D. S., Goldbaum, M., Cai, W., Valentim, C. C., Liang, H., Baxter, S. L., & Zhang, K. (2018). Identifying medical diagnoses and treatable diseases by image-based deep learning. *Cell, 172*(5), 1122–1131.

Kumar, I., Jha, R. S., Kumar, S., & Borah, S. (2014). Brain abnormality detection from MR images using matrix symmetry method. *International Journal of Computer Applications (IJCA) Special Issue on Bioinformatics and Computational Biology,* (3), 22–25. https://www.ijcaonline.org/proceedings/icbcb/number3/17631-1505 (accessed on 11th July 2025).

Latif, G., Butt, M. M., Khan, A. H., Butt, M. O., & Al-Asad, J. F. (2017). Automatic multimodal brain image classification using MLP and 3D glioma tumor reconstruction. In *2017 9th IEEE-GCC Conference and Exhibition (GCCCE)* (pp. 1–9). IEEE.

Mohsen, H., El-Dahshan, E. S. A., El-Horbaty, E. S. M., & Salem, A. B. M. (2018). Classification using deep learning neural networks for brain tumors. *Future Computing and Informatics Journal, 3*(1), 68–71.

Nagle-McNaughton, T. P., Scuderi, L. A., & Erickson, N. (2022). Squeezing data from a rock: Machine learning for Martian science. *Geosciences, 12*(6), 248. https://www.mdpi.com/2076-3263/12/6/248 (accessed on 11th July 2025).

Nichat, A. M., & Ladhake, S. A. (2016). Brain tumor segmentation and classification using modified FCM and SVM classifier. *International Journal of Advanced Research in Computer and Communication Engineering, 5*(4), 73–76.

Ohgaki, H., & Kleihues, P. (2005). Population-based studies on incidence, survival rates, and genetic alterations in astrocytic and oligodendroglia gliomas. *Journal of Neuropathology & Experimental Neurology, 64*(6), 479–489.

Ramitha, M. A., Mohanasundaram, N., & Santhosh, R. (2023). Modelling a fused deep network model for pneumonia prediction. *Computer Systems Science & Engineering, 46*(3). https://www.techscience.com/csse/v46n3/51919 (accessed on 11th July 2025).

Siddique, N., Paheding, S., Elkin, C. P., & Devabhaktuni, V. (2021). U-net and its variants for medical image segmentation: A review of theory and applications. *IEEE Access, 9*, 82031–82057.

Wang, W., Jiang, B., Sun, H., Ru, X., Sun, D., Wang, L., & Feigin, V. L. (2017). Prevalence, incidence, and mortality of stroke in China: Results from a nationwide population-based survey of 480 687 adults. *Circulation, 135*(8), 759–771.

Zhang, J., Jiang, Z., Dong, J., Hou, Y., & Liu, B. (2020). Attention gate resU-Net for automatic MRI brain tumor segmentation. *IEEE Access, 8*, 58533–58545.

CHAPTER 2

Identification of Non-Proliferative Diabetic Retinopathy Stages Using the CLAHE Method of Data Preprocessing

AJITA ARVIND MAHAPADI,[1,*] VISHAL SHIRSATH,[1] and RAJ GAURAV MISHRA[2]

[1]*Department of Computer Engineering, Ajeenkya D. Y. Patil University, Lohegaon, Pune, Maharashtra, India*

[2]*Dr. Vishwanath Karad MIT World Peace University, Pune, Maharashtra, India*

*Corresponding author

ABSTRACT

The most common illness in India is reportedly diabetic retinopathy (DR), in which clogged retinal blood vessels result in symptoms ranging from impaired vision to sight loss. However, effective remedies are available. More than 50% of cases of blindness can be avoided with proper diagnosis, prompt treatment, and the necessary techniques. DR can be the outcome of any kind of diabetes, including gestational diabetes, type 1, and prediabetes. The longer a person has diabetes, the higher their risk. DR will affect more than half of those with the disease. The three main stages of this disease are proliferative DR, non-proliferative DR, and no DR. Ophthalmologists commonly use fundus photography photos to recognize stages in their patients. As a result, there is a significant imbalance between the different types of DR in the fundus image collection. This research paper provides an overview for classifying different DR phases using DL and ML techniques.

Intelligent Computing and Analytics in Image Processing. Samarjeet Borah, Jambi Ratna Raja Kumar, Laxmi Bewoor, Rahul Raghvendra Joshi, and Mahendra Prabhakar Deore (Eds.)

DOI: 10.1201/9781779640772-2

2.1 INTRODUCTION

One of the chronic conditions that might cause a person's mortality is diabetes. Furthermore, according to the WHO (2023), there were more than 400 million cases of diabetes in 2014, up from 108 million cases in 1980. Comparative analysis for DR medical conditions that lead to blindness has been done; from this analysis, it has been confirmed that DR has been increasing in Small and Mid-income nations compared to High-Income countries. Risk detection identification is one issue that can lessen the number of people needing treatment. According to recommendations for societal impact, DR stage classification can aid society in preventing eyesight loss in DR patients. If DR can be identified early, a person can avoid reaching the PDR stage, leading to blindness in people. This research aims to identify stages and help avoid blindness if it is identified in initial stage. Compared to the economic impact, diabetes can cause economic loss due to the expense of medications, which can affect a person's economic loss due to ongoing blood sugar monitoring, medications, the discovery of retinal disorders, or other things. Additionally, compared to other nations, people's economic situation is less favorable in countries like India. Thus, there is a financial burden on individuals, especially in countries like India, where the economic situation is less favorable compared to other nations.

2.1.1 TYPES OF DIABETES

1. **Type 1:** When a person's immune system cannot produce insulin-generating cells, this condition leads to type 1 diabetes. More than 5% (https://www.cdc.gov/diabetes/basics/diabetes.html) of diabetic patients have type 1 diabetes. This stage of diabetes can occur at any age but is often observed in young adults and children.
2. **Type 2:** In this form, the body either does not produce enough insulin (https://www.cdc.gov/diabetes/basics/diabetes.html) or its cells do not respond appropriately. This type of diabetes is the most prevalent. Although it primarily affects adults, it can also affect children.
3. **Prediabetes:** Blood sugar levels are higher than the normal range but not high enough to be diagnosed as diabetes.

4. **Gestational:** This type occurs in pregnant women and generally resolves after childbirth. The risk of developing type 2 diabetes increases if a woman has gestational diabetes.

The optic nerve controls your vision. It is an extension of your spine and brain's neural systems. Your eyes transfer impulses to your brain using the optic nerve. Your brain interprets this sensory data so that you can see. The area of your eye known as the *macula* is responsible for processing your central vision, or what you see directly in front of you. It is essential to your vision. This region is the rounded portion of your retina in the center, at the rear side of your eye. The *fovea centralis*, a tiny depression within the neurosensory retina, has the highest possible visual acuity score—a measure of the eyes' capacity to detect various kinds of information about objects at a given distance. The center of this structure is where the *fovea* of the *macula*, which regulates central vision, is located.

An aneurysm is a swelling that emerges in an artery in the bloodstream due to a narrowing in the outermost layer of the vessel, generally where it branches. When fluid passes through a blood vessel that has weakened, high blood pressure (BP) allows a tiny portion to expand outward. Hard exudates of the retina are commonly seen in pupils with diabetic retinopathy (DR). Albumin and fibrinogen are examples of lipid and proteinaceous molecules that have the potential to enter the hard exudates resulting from blood-retinal barrier disturbances. They are frequently deposited in the plexiform layer located on the peripheral retina. Hard exudates represent distinctive intra-retinal yellow-white deposits that can appear as specks or extensive areas and later expand into circinate rings.

The light-sensitive tissue and the retina are found on the rear wall of the eye and bleed whenever the individual has retinal hemorrhage (RH), also known as bleeding in the retina.

2.1.2 STAGES OF NON-PROLIFERATIVE DIABETIC RETINOPATHY (NPDR)

A progressive eye condition known as diabetic retinopathy (DR) has two forms and four phases. Non-proliferative and proliferative are the two categories. Early disease phases are referred to as non-proliferative, while advanced disease stages are referred to as proliferative.

2.1.2.1 Mild

In early stages of DR, the walls of the retinal blood vessels weaken. Slight lumps that frequently leak blood into the retina project from the vessel walls. The retina's tissues could enlarge, resulting in white patches.

2.1.2.2 Moderate

Aside from hard exudates, cotton wool patches, and venous beading, some patients also have hemorrhages or macular holes in one to three retinal quadrants.

2.1.2.3 Severe

The blood flow in the retina experiences a notable reduction due to a higher degree of obstruction in its blood vessels. The regulatory board has granted authorization for the initiation of angiogenesis in the retinal region.

2.2 LITERATURE REVIEW

Cahoon et al. (2022) presented that detection and diagnosis are essential for managing the indications and possibly slowing the course of these diseases. Ophthalmologists employ fundus photography as a cost-effective and reliable diagnostic tool. Fundus photography datasets, however, experience a severe disparity between the various DR classes. The arrangement of DR into one of three groups—mild, moderate, and severe—by a two-stage deep convolutional neural network (CNN) (Cahoon et al., 2022) structural design is accomplished in this paper using a fine-tuned ResNet-18. Further classification of moderate DR was performed using a fine-tuned ResNet-50, while classification of severe DR was also accomplished with a fine-tuned ResNet-50. The proposed design uses a preprocessing stage that includes data augmentation and image scaling.

In the study conducted by Jena et al. (2023), it was observed that researchers employed the CLAHE technique (Jena et al., 2023) specifically on the green spectrum of the image. This approach was adopted due to the enhanced visibility of bright lesions, which are more discernible in this particular channel. The present research proposed a unique unsymmetrical deep learning (DL)-based characteristics approach for identifying the presence of DR. The optic disc and blood vessels were segmented using asymmetry DL features extracted via U-Net. Then, an SVM and a CNN were applied to identify DR lesions. Hemorrhages, microaneurysms, exudates, and regular lesions are the four categories for separating the lesions. The proposed method has been tested using two publicly available retinal image datasets: MESSIDOR and APTOS. The specificity of the results for non-DR detection for the APTOS and MESSIDOR datasets is 98.6% and 91.9%, respectively. The exudate diagnosis accuracy rates over these two datasets are 96.9% and 98.3%, respectively.

With dark- and light-colored lesions, Rajput et al. (2022) proposed an algorithm for identifying and characterizing lesions linked to non-proliferative DR. The diagnosis of bright and dark disorders can be achieved through the use of a multilayered perceptron, as described in the reference. This approach relies on the identification of exudates and cotton wool patches for the diagnosis of bright disorders, whereas microaneurysms and hemorrhages are considered indicative of dark disorders. Instead of reviewing a primary categorization method for DR lesions, the multilayered perceptron algorithm—a technique with an accuracy score of 0.99 and a true positive rate (TP Rate) of 96%—is discussed.

According to Bidwai et al. (2022), DR is a highly widespread cause of visual impairment on a global scale. This condition arises as a result of prolonged diabetes accompanied by inconsistent levels of blood glucose. The prevalence of this issue among individuals of working age necessitates prompt attention in order to mitigate the risk of potential visual impairment in the future. In the preliminary phase, diagnostic and evaluative measures for DR have been implemented using artificial intelligence (AI)-based technologies. The timely identification of visual impairments enables the implementation of appropriate interventions, thereby mitigating the development of ocular complications. The current review explores many techniques for the detection of DR using indicators such as blood vessels, microaneurysms, exudates, macula, optic discs, and hemorrhages. The majority of studies utilize fundus images of the retina obtained through the use of a fundus camera. According to the author, this literature has yet to examine new cutting-edge techniques with the PRISMA (Bidwai et al., 2022) methodology and AI at its foundation.

Chang et al. (2022) states that the fundamental goal of this study is to deploy various ML algorithms to indicate diabetes diagnosis. Furthermore, these models are compared to establish the most effective one by measuring their prediction accuracy and other performance metrics, including precision, recall, and F1 score. Random Forest (Chang et al., 2022) outperformed the other models studied, with an accuracy of 82.26%.

The individuals mentioned include Tokuda Yoshihiro, Hitoshi Tabuchi, and a reference denoted by Tokuda et al. (2022). An automated fundus camera was employed to exclusively capture RHs, while a deep CNN, a ML algorithm, was utilized for the detection and classification of DR. The study focused on detecting moderate-to-worse non-proliferative diabetic retinopathy (NPDR) as well as classifying mild-to-worse NPDR. The area under the curve (AUC) results for determining the presence of mild-or-worse DR were 0.812, 0.888, and 0.884, respectively, whereas those for detecting moderate-or-worse DR were 1.0, 1.0, and 1.0. The study's automated AI-based measurement, which considers only RH, may be used to recognize DR that requires ophthalmologist intervention.

According to a biological system formulated by Abdelsalam and Zahran (2021), the retinal microvascular network has unusual multifractal elements involving generalized dimensions, lacunarity, and singularity spectrum. This work provides a comprehensive overview of a unique multifractal geometry-based technique for early detection of DR. Early NPDR can be recognized by analyzing images from macular optical coherence tomography angiography (Abdelsalam and Zahran, 2021), automating the problem-solving process, and improving precision using supervised ML algorithms like the SVM algorithm. The classification technique achieved an accuracy of 98.5%.

The findings presented in the study conducted by Liu (2018) demonstrate the clinical relevance of automated approaches for detecting DR, thereby highlighting their importance. This paper proposes a new method for detecting microaneurysms and hard exudates in DR using a deep symmetric CNN (Liu et al., 2016). The incorporation of a symmetric convolutional structure significantly improves the efficacy of feature extraction. The proposed technique also addresses the issue of imbalanced positive and negative samples to mitigate overfitting by expanding and deepening the network. The detection accuracies for the items are 93.6%, 93.2%, and 92.0% while employing different processing structures—namely, average pooling, max pooling, and convolution, respectively.

Sunil and Anusuya (2021) presents control mechanisms for recognizing various traits and variations in the color retina using multiple categories. Recovery from diabetes is described by this therapeutic strategy (Sunil and Anusuya, 2021) using data from several datasets, including DRIDB0, DRIDB1, MESSIDOR, STARE, and HRF. Issues related to ML, neural networks, and DL algorithms are discussed, along with associated metrics such as sensitivity, precision, accuracy, specificity, and F1 score, as well as comparisons involving the Matthews correlation coefficient (MCC) and kappa coefficient. Finally, compared to other methods, the findings were more successful due to the DL strategy. The technology can aid ophthalmologists in the early detection of diabetic signs, improve treatment, and enhance biological quality of life.

Ebin and Ranjana (2020) proposed that the most effective way to preserve vision is through early identification of DR. This article refers to current methodologies and studies employing patient fundus images to diagnose DR in its initial stages. In this work, the early stages of automated detection of DR are examined using patient fundus images and VGG and Inception V3 (Ebin and Ranjana, 2020). An assessment of DR-accessible datasets for color fundus images of the retina was also conducted.

Anumol, Anamika, and Kurian (2022) implemented an automated DR screening approach designed to accelerate identification and decision-making processes, aiding in the management or control of DR progression. Using ML models such as VGG-16, CNN, and VGG-19, this work proposes an automated classification system that evaluates fundus images with varying illumination and visual fields, assigning a severity score for DR.

Lifeng, Zhu, and Zhou (2020) reported that macular edema and microaneurysms in the retina are the first symptoms of DR, and quick diagnosis can reduce the risk of non-proliferative DR. DL is rapidly advancing, making it an increasingly effective method for addressing medical image analysis challenges. The proposed system analyzes whether a microaneurysm is present in a fundus image using CNN algorithms that incorporate DL as a core component, accelerated by graphics processing units. The semantic segmentation technique classifies the fundus image as normal or diseased. To identify the distinctive features of a microaneurysm, semantic segmentation is used to partition image pixels based on shared semantic meaning.

Tieyuan et al. (2021) suggested that the therapeutic importance of adopting autonomous technologies to identify DR in its early stages cannot be overstated. This study proposes a novel detection approach based on deep symmetric CNN for identifying microaneurysms and hard exudates of DR. The efficiency of feature extraction is increased by using the symmetric convolutional structure. The asymmetric convolutional structure also improves the efficiency of feature extraction. The proposed approach aims to mitigate the issue of overfitting by expanding and enhancing the system, thereby addressing the disparity between positive and negative data. Furthermore, a variety of network topologies are employed to apply diverse feature filters during the feature extraction process.

Xiaoliang et al. (2018) proposed that DL techniques using CNNs have recently emerged as a viable method for analyzing biological images. According to ophthalmologists' knowledge, representative DR photographs in this study have been clustered into five groups. Inception Net V3 has demonstrated the effectiveness of using deep CNNs for DR image identification by achieving state-of-the-art accuracy.

Zeynab et al. (2019) implemented a method to autonomously distinguish grades 0, 1, 2, and 3 of DR from fundus images. A technique based on the visibility graph and Radon transform was presented. The two steps of the suggested technique are feature extraction and classification. The VG approach was used for feature extraction in image processing. Then, these features were added to the error-correcting output codes (ECOC) approach for classification purposes.

Jagadish et al. (2008) proposed using fundus images to diagnose the stage of DR through a computer-based method. In the context of fundus images, several image preprocessing techniques, morphological processing methods, and texture analysis approaches are employed to detect and characterize certain features, such as the precise localization of hard exudates, blood vessels, and contrast variations.

According to Peng et al. (2021), a prominent contributor to avoidable cases of juvenile visual impairment is a retinal condition known as retinopathy of prematurity (ROP), which primarily affects infants born prematurely or with low birth weight. This study proposes a novel and efficient deep neural network architecture for the purpose of 5-level ROP staging. The network comprises a concatenation-based deep feature fuser, a multistream parallel feature extractor, and an ordinal classifier that aligns with the clinical 5-level ROP staging. The initial feature extractor is a three-stream parallel architecture comprising EfficientNetB2, DenseNet121, and ResNet18 models, which are capable of extracting diverse and rich advanced features. Furthermore, the use of concatenation and convolution techniques is essential in integrating information from the three streams, resulting in a more valuable and comprehensive output.

Walter (2002) reports that exudates are identified by their high degree of gray-level variation, and their outlines are established using morphological reconstruction methods. For this method to work, the optic disc must be detected. The optic disc can be identified using morphological filtering methods and the watershed transformation. Using a small image database, the algorithm's performance was evaluated against that of a human grader (Table 2.1).

TABLE 2.1 Comparative Analysis

Sl. No.	Name of Author	Dataset	Technique	Results
1.	Cahoon (2022)	Kaggle APTOS	5-fold cross-validation and 2-stage deep convolutional neural network	91% in the first stage; 90% and 80% in the second sub-stages
2.	Jena (2023)	APTOS and MESSIDOR	CNN with SVM, U-Net for segmentation	91.9% for MESSIDOR and 98.6% for APTOS
3.	Rajput (2022)	Collected from Neha Deshpande and Manoj Saswade	Multilayered perceptron	True positive rate: 96%, false positive rate: 0%, accuracy: 0.99
4.	Bidwai (2022)	–	PRISMA approach and artificial intelligence	78 research studies
5.	Chang (2022)	RFSS	Random forest	Accuracy: 82.26%
6.	Tokuda et al. (2022)	Department of Diabetology, Tsukazaki Hospital (Himeji City, Japan)	Deep convolutional neural network	AI-based automatic diagnosis focusing on RH for DR detection
7.	Abdelsalam (2021)	Ophthalmology Center, Mansoura University, Egypt	Support vector machine	Accuracy: 98.5%
8.	Liu (2021)	DIARETDB1 (DB1)	Deep symmetric convolutional neural network	Convolution: 92.0%, max pooling: 93.2%, average pooling: 93.6%
9.	Sunil (2021)	DRIDB0, DRIDB1, MESSIDOR, STARE, and HRF	Therapeutic approach, deep learning	More effective results
10.	Ebin (2020)	MESSIDOR, DIARETDB	Transfer learning	Reuse learning techniques

2.3 DATA PREPROCESSING

By reorganizing the pixel intensities to maximize the average intensity and detail, the image processing technique known as histogram equalization can improve the brightness of an image. This method alters the frequency distribution of an image's pixel values to produce a more uniform histogram. A uniform histogram ensures that each pixel in the image has an equal probability of appearing, thereby boosting contrast and distributing brightness more evenly across the image.

The histogram equalization variant known as CLAHE is frequently used in image processing applications to enhance image contrast while avoiding excessive noise amplification and preserving characteristic features. The primary objective of CLAHE is to limit histogram equalization so that it is applied only to specific regions of the image, rather than the entire image. Contrast enhancement and contrast limitation are the two critical stages of CLAHE. The first stage enhances contrast by performing localized histogram equalization in each image region. In the next stage, a non-linear function is applied to the histogram to reduce the number of pixels with extremely high or low intensities, thereby minimizing excessive contrast in the image. The clip limit parameter, which determines the amount of contrast applied, is used to generate the non-linear function (Figures 2.1–2.3).

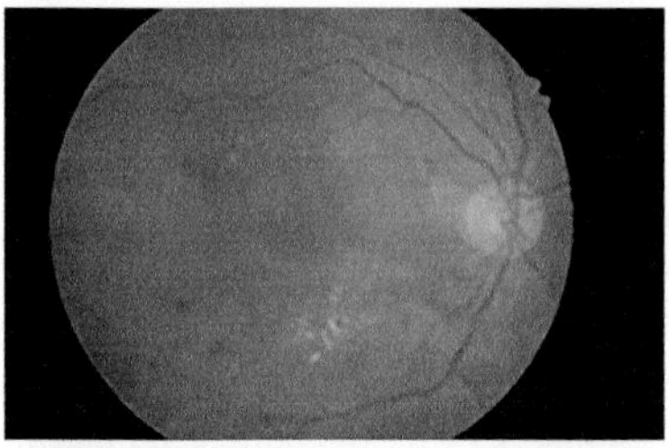

FIGURE 2.1 Original fundus image.

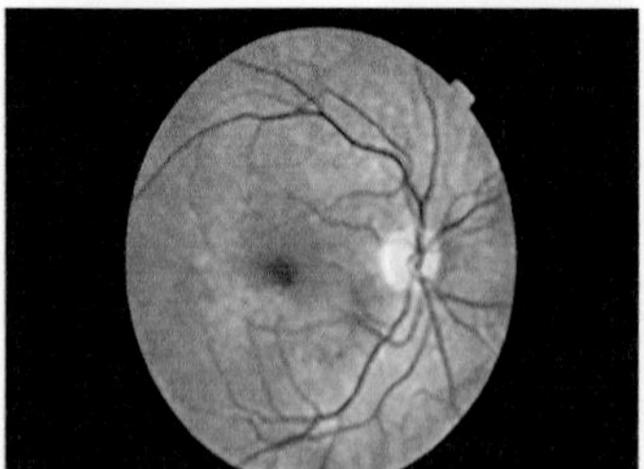

FIGURE 2.2 Grayscale image after application of the CLAHE method.

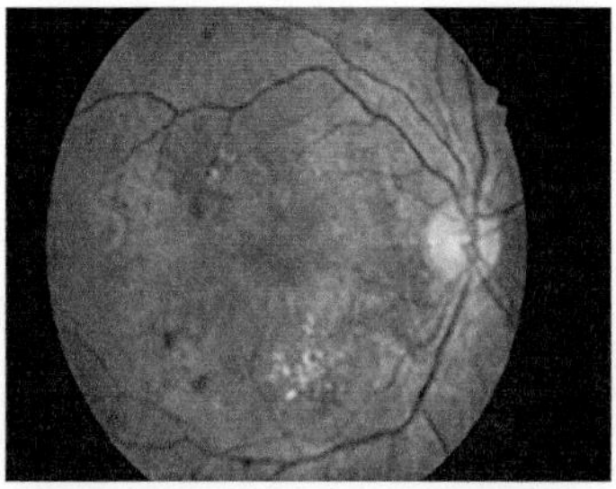

FIGURE 2.3 Colored image after applying the CLAHE method.

From Figures 2.1–2.3, we can identify the difference between the two fundus images. In Figure 2.1, the original fundus image is shown, whereas after applying the CLAHE data preprocessing method, the features of the images become more visible. Grayscale output is used to gain more visibility of hard exudates, microaneurysms, and blood vessels. Colored image output distinguishes hard exudates, where yellow-white intra-retinal deposits can be identified. A histogram presentation of both input and output fundus images is shown below, where a graph is plotted between pixel values and the frequency of the fundus image. In Figure 2.4, the distribution is uneven, but the frequency distribution of pixel values is even in Figure 2.5.

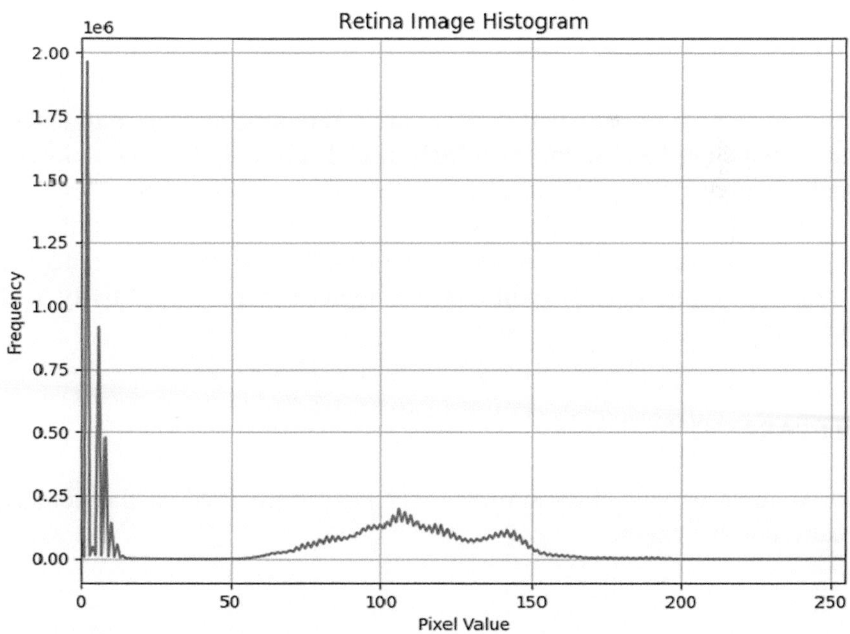

FIGURE 2.4 Histogram of fundus image.

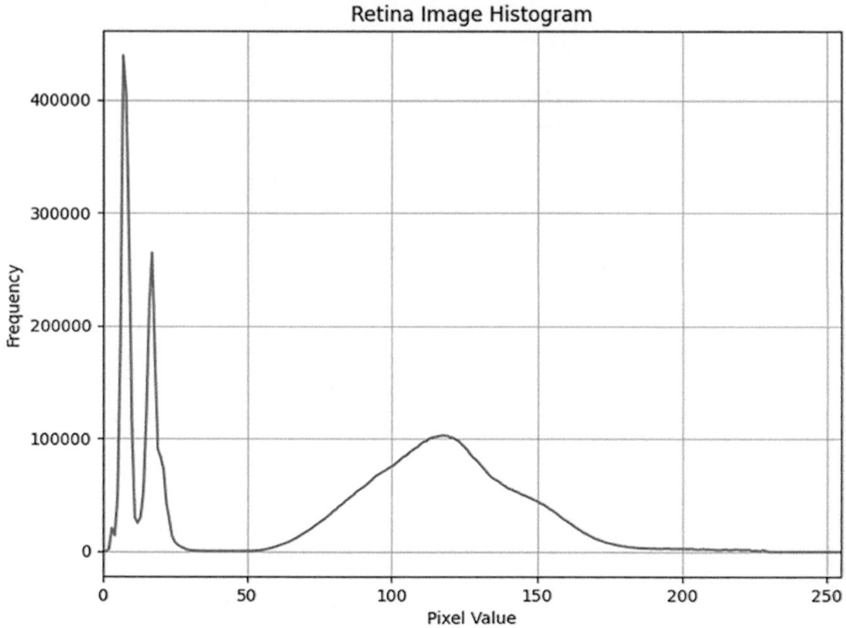

FIGURE 2.5 Histogram of fundus image after applying CLAHE.

2.4 CONCLUSION

The classification of DR based on stages can be done as no DR, NPDR (mild, moderate, severe), and PDR. Fundus photography is used to detect stages, which ophthalmologists widely use. Therefore, the fundus image database undergoes extreme disparity among different classes of DR. This research proposal will propose a novel method to classify different stages of DR using various image enhancement techniques. The outcome obtained from the datasets will achieve higher accuracy, sensitivity, and specificity.

DECLARATION

No AI tools or any other similar tools have been used in the content creation of this chapter.

KEYWORDS

- **convolution neural network**
- **deep learning**
- **diabetic retinopathy**
- **error-correcting output codes**
- **fundus images**
- **machine learning**
- **microaneurysms**
- **non-proliferative diabetic retinopathy**
- **random forest**

REFERENCES

Abdelsalam, M. M., & Zahran, M. A. (2021). A novel approach of diabetic retinopathy early detection based on multifractal geometry analysis for OCTA macular images using support vector machine. *IEEE Access, 9*, 22844–22858. https://doi.org/10.1109/ACCESS.2021.3054743 (accessed on 11th July 2025).

Bidwai, P., Gite, S., Pahuja, K., & Kotecha, K. (2022). A systematic literature review on diabetic retinopathy using an artificial intelligence approach. *Big Data and Cognitive Computing, 6*(4), 152. https://doi.org/10.3390/bdcc6040152 (accessed on 11th July 2025).

Cahoon, S., Shaban, M., Switala, A., Mahmoud, A., & El-Baz, A. (2022). Diabetic retinopathy screening using a two-stage deep convolutional neural network trained on an extremely unbalanced dataset. *SoutheastCon 2022, Mobile, AL, USA*, 250–254. https://doi.org/10.1109/SoutheastCon48659.2022.9764079 (accessed on 11th July 2025).

Centers for Disease Control and Prevention (CDC). https://www.cdc.gov/diabetes/basics/diabetes.html (accessed on 11th July 2025).

Changa, V., Ganatrab, M. A., Hallb, K., Golightlya, L., & Xua, Q. A. (2022). An assessment of machine learning models and algorithms for early prediction and diagnosis of diabetes using health indicators. *Healthcare Analytics, 2*, 100118. https://doi.org/10.1016/j.health.2022.100118 (accessed on 11th July 2025).

Ebin, P., & Ranjana. (2020). An approach using transfer learning to disclose diabetic retinopathy in early stage. In *International Conference on Futuristic Technologies in Control Systems & Renewable Energy (ICFCR). IEEE Xplore*. https://doi.org/10.1109/ICFCR50903.2020.9249988 (accessed on 11th July 2025).

GeeksforGeeks. https://www.geeksforgeeks.org/introduction-deep-learning (accessed on 11th July 2025).

JavaTpoint. https://www.javatpoint.com/machine-learning-techniques (accessed on 11th July 2025).

Jena, P. K., Khuntia, B., Palai, C., Nayak, M., Mishra, T. K., & Mohanty, S. N. (2023). A novel approach for diabetic retinopathy screening using asymmetric deep learning features. *Big Data and Cognitive Computing, 7*. https://doi.org/10.3390/bdcc7010025 (accessed on 11th July 2025).

Liu, T., Chen, V., Shen, H., Zhon, R., Zhang, M., Tonglailiu, & Liu, J. (2021). A novel diabetic retinopathy detection approach based on deep symmetric convolutional neural network. *IEEE Access, 9*, 160552–160558. https://doi.org/10.1109/ACCESS.2021.3131630 (accessed on 11th July 2025).

Liu, T., Chen, Y., Shen, H., Zhon, R., Zhang, M., Tonglailiu, & Liu, J. (2016). A novel diabetic retinopathy detection approach based on deep symmetric convolutional neural network. *IEEE Access, 4*. https://doi.org/10.1109/ACCESS.2021.3131630 (accessed on 11th July 2025).

Mohammadpoury, Z., Nasrolahzadeh, M., Mahmoodian, N., & Haddadnia, J. (2019). Automatic identification of diabetic retinopathy stages by using fundus images and visibility graph method. *Measurement, 140*, 133–141. https://doi.org/10.1016/j.measurement.2019.02.089 (accessed on 11th July 2025).

Nayak, J., Bhat, P. S., Acharya, R. U., Lim, C. M., & Kagathi, M. (2008). Automated identification of diabetic retinopathy stages using digital fundus images. *Springer*. https://doi.org/10.1007/s10916-007-9113-9 (accessed on 11th July 2025).

Peng, Y., Zhu, W., Chen, Z., Wang, M., Geng, L., Yu, K., Zhou, Y., Wang, T., Xiang, D., Chen, F., & Chen, X. (2021). Automatic staging for retinopathy of prematurity with deep feature fusion and ordinal classification strategy. *IEEE Transactions on Medical Imaging, 40*(7), 1–10.

Qiao, L., Zhu, Y., & Zhou, H. (2020). Diabetic retinopathy detection using prognosis of microaneurysm and early diagnosis system for non-proliferative diabetic retinopathy based on deep learning algorithms. *IEEE Access, 8*. https://doi.org/10.1109/ACCESS.2020.2993937 (accessed on 11th July 2025).

Rajput, Y., Gaikwad, S., Dhumal, R., & Gaikwad, J. (2022). A novel method for non-proliferative diabetic retinopathy disease classification using image processing techniques. *Bornova Izmir, 20*(11), 8002–8011. https://doi.org/10.14704/nq.2022.20.11.NQ66794 (accessed on 11th July 2025).

Sajan, A., Anamika, K., & Kurian, S. M. (2022). Diabetic retinopathy detection using deep learning. *International Journal of Engineering Research & Technology (IJERT), 10*(4, Special Issue – 2022), 154.

Sunil, S. S., & Anusuya, S. (2021). A novel system for the identification of diabetic retinopathy using computer-based statistical classification. *Turkish Journal of Computer and Mathematics Education, 12*(3), 4044–4059.

Tokuda, Y., Tabuchi, H., Nagasawa, T., Tanabe, M., Deguchi, H., Yoshizumi, Y., Ohara, Z., & Takahashi, H. (2022). Automatic diagnosis of diabetic retinopathy stage focusing exclusively on retinal hemorrhage. *Medicina, 58*(11), 1681. https://doi.org/10.3390/medicina58111681 (accessed on 11th July 2025).

Walter, T., Klein, J. C., Massin, P., & Erginay, A. (2002). A contribution of image processing to the diagnosis of diabetic retinopathy: Detection of exudates in color fundus images of the human retina. *IEEE Transactions on Medical Imaging, 21*(10), 1236–1243. https://doi.org/10.1109/TMI.2002.806290 (accessed on 11th July 2025).

Wang, X., Lu, Y., Wang, Y., & Chen, W. B. (2018). Diabetic retinopathy stage classification using convolutional neural networks. In *IEEE International Conference on Information Reuse and Integration for Data Science*. https://doi.org/10.1109/IRI.2018.00074 (accessed on 11th July 2025).

World Health Organization (WHO). (2023). https://www.who.int/health-topics/diabetes#tab=tab_1 (accessed on 11th July 2025).

CHAPTER 3

Optimized Deep Learning Model for Predicting Brain Tumors

NARAYAN B. VIKHE[1*] and MANISH SHRIVASTAVA[2]

[1]*Department of Computer Engineering, Dr. Vithalrao Vikhe Patil College of Engineering, Ahilyanagar, Maharashtra, India*

[2]*Department of Computer Science and Engineering, Vivekananda Global University, Jaipur, Rajasthan, India*

*Corresponding author

ABSTRACT

First, scanning manually through a huge amount of MRI images to detect a brain tumor is incredibly time-consuming and often inaccurate in modern times. This may delay providing the patient with adequate medical treatment. Additionally, it can be a very lengthy process because it requires collecting a large number of images. It is difficult to distinguish tumor regions since brain tumor cells and normal tissue have somewhat similar appearances. Consequently, there is a need for a highly accurate automated tumor detection system. This chapter proposes an approach to segment 2D MRI brain scans that show brain tumors using a convolutional neural network (CNN). Deep learning methods and traditional classifiers follow. We collected several MRI images with various tumor sizes, locations, and image resolutions in order to train the model effectively. We incorporated the support vector machine classifier and several activation techniques to validate our findings. Python is a powerful programming language for rapid development; thus, we used TensorFlow and Keras in it. Due to our work, the CNN outperformed the existing state of the art, achieving an accuracy of 98.74%. In conclusion, by assisting doctors in

Intelligent Computing and Analytics in Image Processing. Samarjeet Borah, Jambi Ratna Raja Kumar, Laxmi Bewoor, Rahul Raghvendra Joshi, and Mahendra Prabhakar Deore (Eds.)

DOI: 10.1201/9781779640772-3

precisely identifying brain tumors using MRI, our deep CNN-based model significantly expedites treatment.

3.1 INTRODUCTION

Several minimally invasive methods for peering inside the body are referred to as "medical imaging" (Kasban, Elbendary, and Salama, 2015). Medical imaging is largely employed for therapeutic and diagnostic purposes in the human body. As a result, it is crucial for enhancing care and human health. Image segmentation is an important step in image processing that must be successful for higher-level image processing to function properly (Maxwell, 1892). In this instance, we mainly concentrated on segmenting brain tumors using an MRI scan. This simplifies the process for health professionals to identify where the brain tumor is located. Medical imaging is the use and study of 3D body models in digital form, typically acquired by CT scan or MRI, to diagnose illness or provide immediate medical assistance such as surgery or surgical planning. Medical imaging is used by physicians, engineers, and radiologists to thoroughly comprehend the structure of specific patients or populations. A more thorough comprehension of the interplay between the human skeleton and medical equipment is possible using statistical evaluation and measurement, and the creation of simulation models that accurately depict societal anatomical shapes. The term "tumor" (American English) is synonymous with "tumour" (British English), which is caused by abnormal cell proliferation. Cancer and tumors differ significantly from one another (Surya and Satheesh, 2016).

3.1.1 CLASSIFICATION OF TUMORS

3.1.1.1 Benign tumor

Not all benign tumors are malignant or cancerous. They might not invade adjacent tissues or spread to other parts of the body, unlike cancer. Benign tumors usually have a very good prognosis, but if they affect vital organs like blood vessels or neurons, complications can become severe.

3.1.1.2 Malignant tumor

Malignant tumors are a specific kind of cancer that arises when cancer cells proliferate uncontrollably. The cancer advances to the next stage as the cancer cells continue to divide and multiply. Malignant tumors spread quickly to other parts of the body and grow rapidly.

The most recent study from 2021 (Oelze, Zachary, and O'Brien, 2003) predicts that 24,630 people (13,890 men and 10,730 women) would receive a diagnosis of brain and spinal cord cancer in the country. Just over 1% of people will develop this kind of brain tumor in their lifetime. It accounts for between 80 and 85% of all primary cancers of the central nervous system. A total of 3460 children under the age of 14 will receive a diagnosis of a brain tumor in 2021, in addition to primary brain tumors. Brain and other central nervous system cancers are the tenth most common cancers, affecting both men and women. A total of 18,600 people may die from primary brain and central nervous system cancer in 2021. Consequently, improving the precision of previously proposed methods is essential to enhancing medical imaging. In our study, a deep CNN-based algorithm with an accuracy of 98.74% is suggested to aid doctors in their clinical work by eliminating the need for manual examination of MRI images and accelerating treatment.

3.2 METHODS OF SEGMENTATION OF BRAIN TUMORS

Methods for segmenting brain tumors can be categorized into three groups: completely automatic, semi-automatic, and manual. These methods can be customized based on the required level of user participation (Cancer Net Editorial Board, 2021).

3.2.1 MANUAL SEGMENTATION METHODS

The professional must combine the diverse data from MRIs with the anatomical and physiological understanding they have gained through education and practice. The doctor must carefully isolate the tumor regions by hand after examining the brain tumor in multiple sections of the images. This requires a lengthy period, as there is substantial intra- and inter-rater variability with manual segmentation, which depends on the physician (Gordillo, Montseny, and Sobrevilla, 2013). Despite being frequently utilized, manual segmentation cannot replace completely automatic or semi-automatic procedures.

3.2.2 SEMI-AUTOMATIC TECHNIQUES FOR SEGMENTATION

User input is necessary for the three fundamental functions of initiation, involvement, and assessment (White et al., 1999). The major starting step is defining a problem area, which restricts the calculated tumor region that the automated algorithm processes. Additionally, the preprocessing technology's settings may be adjusted to match the input images. Automated algorithms can be guided toward the intended outcome via feedback during the process, in addition to the initial setup. Feedback-based modifications are also made during this procedure. If the user is dissatisfied with the results, they can adjust or repeat the process.

A method of segmenting tumors was proposed by Hamachi et al. (2012). Using this approach, the algorithm was applied independently to each MRI modality.

A novel classification approach was applied to the existing semi-automatic method (Hamamci, 2012). Using this method, the segmentation challenge is transformed into a classification problem, and the brain tumor is segmented within the same brain based on training and classification. Several brain MRI images (with varying responses) taken at different times are often required for classification using a machine learning (ML) approach for brain tumor categorization. The output must address significant distortion correction and additional noise. However, using this approach, the user conducts the operation by classifying a selection of vectors related to each tissue type in a single example. The support vector machine (SVM), which is used to classify all vectors in an image according to their respective tissue types, is trained by extracting intensity values and spatial references of subsets of these vectors as features.

While a semi-automated technique for brain tumor segmentation can achieve good outcomes in less time than a manual approach, variability between users and evaluators still exists. Because of this, most recent research on brain tumor segmentation relies on fully automated methods.

3.2.3 ABSOLUTE AUTOMATED SEGMENTATION TECHNIQUES

No user engagement is necessary with this method. In particular, the segmentation issue is resolved by combining domain expertise with artificial intelligence (AI). Automated techniques address the segmentation of gliomas, a type of tumor that develops in the brain. Clinical brain cancer scans or artificial databases' MRI data (Havaei et al., 2016) are inherently complex.

Because MRI equipment and acquisition processes vary from scan to scan, each distinct region of the image in the dataset may show variations in quality and other factors. The need to segment tumor regions using multiple procedures adds further complexity.

3.2.4 BRATS DATASET

BRATS stands for benchmark for multimodal brain tumor image segmentation. To be honest, comparing the effectiveness of several cutting-edge techniques for segmenting brain tumor images is a difficult task. Furthermore, utilizing this shared dataset, and using the popular BRATS standard, multiple *glioma* segmentation methods for automated brain cancer segmentation can be objectively compared (Prastawa, Bullitt, and Gerig, 2009). The most recent version (2020) of the BRATS training dataset includes 369 multimodal MRI images for evaluation (Wang, 2020). An assessment tool available exclusively online can be used to evaluate test outcomes. The tool mainly presents results in terms of known Dice scores for sensitivity (positive) and specificity (negative) in three main tumor components: active cells in a tumor, complete tumor (everything related to the tumor), and core tumor (all tumor-related areas excluding edema).

The main use of Dice scoring is in performance measurement. T1 is the actual tumor region in the ground truth for each tumor location, and P1 is the segmented tumor region as determined by the proposed approach.

3.3 THE PROPOSED METHOD

The use of convolutional neural networks (CNNs) in medical imaging is a well-established approach. A CNN is a type of artificial neural network (ANN) specifically designed to understand the components of image analysis and is widely used in image recognition. Deep CNN is a powerful computational image processing method that employs deep learning (DL) to perform productive and detailed tasks. It is commonly used in recommender systems, natural language processing (NLP), image and video recognition, and machine vision.

Similar to the way neurons function in the human brain, hardware and software can work together to form neural networks. However, ANNs are not always the ideal image processing technique. The multilayer architecture used by CNNs helps reduce processing demands. By removing limitations and improving image processing capabilities, CNNs produce systems that are more efficient and better suited to image processing and conversational language tasks.

We enhanced the original CNN model by making several modifications. With fourteen stages and hidden layers in our 9-layer CNN model, we achieved excellent tumor detection performance. The proposed technique is illustrated in Figure 3.1 with a brief description. To standardize dimensions, we applied our proposed method to resize a set of input images to 128 × 128 × 3. Thirty-two 2 × 2 convolutional filters, each supporting three-channel tensors, are managed by a kernel with convolutions created alongside the input layer.

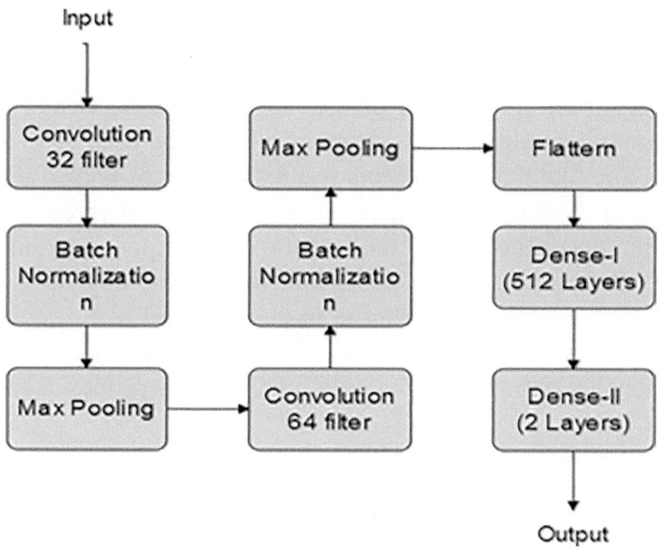

FIGURE 3.1 A proposed method for tumor detection using a deep CNN.

We frequently use the ReLU activation function. ReLU (rectified linear unit) is a piecewise linear operation that outputs the input directly if it is positive. Additionally, we applied the batch normalization method described by Ioffe et al. in 2015. By normalizing layers through re-centering inputs and scaling, batch normalization (Ioffe and Szegedy, 2015) (also known as array norm) not only speeds up neural networks but also increases their stability. It helps accelerate algorithms (*see* Figure 3.1).

Adding the features in the area that the 2D filter covers and swiping it over each channel of the object map are additional steps in the merging process. The dimensions of the output after a dimensioned feature map aggregation layer is $d_h \times d_w \times d_c$:

$$(d_h - fs + 1)/s \times (d_w - fs + 1)/sl \times d_c \tag{1}$$

where; d_h is the height; d_w is the width; d_c is the number of channels; *fs* is the filter size; and *sl* is the step length.

Feature maps can be made smaller by pooling layers. They utilize fewer parameters and therefore reduce the number of calculations carried out within a network. The feature map produced by the convolution layer's extent is gathered by the pooling layer's characteristics. As a result, rather than using the carefully placed characteristics generated by the convolution layer, further actions are carried out using summary features. This makes the model stronger and more resilient to variations in the placement of parts in the input image.

After that, we carried out another round of 64-filter convolution, max pooling, and batch normalization before reduction (Figure 3.2). We constructed two dense layers, where the first layer has 512 units, and the final two units are in the second dense layer. Since softmax gives more precision than the alternatives, we decided to use it as the activation function in the top layer. RMSProp is an extension of surface irregularity descent, and a flexible loss function called categorical_crossentropy was once again used.

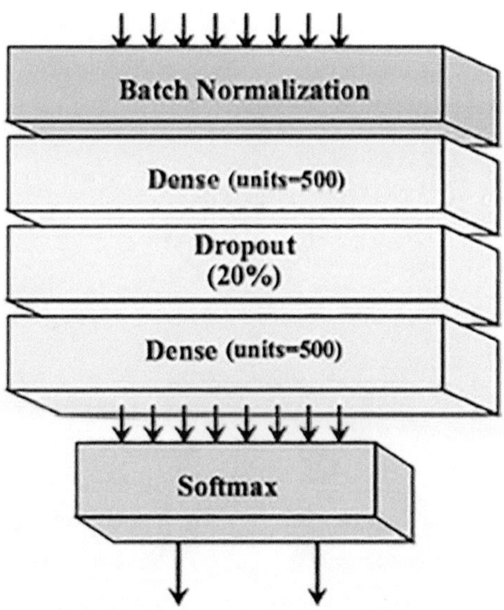

FIGURE 3.2 Batch normalization.

- **Gradient Algorithm:** This optimizer-based form of gradient descent modifies the step size for each parameter by taking a decreasing average of the partial gradients. Figure 3.3 shows the operational flow of the suggested CNN model.

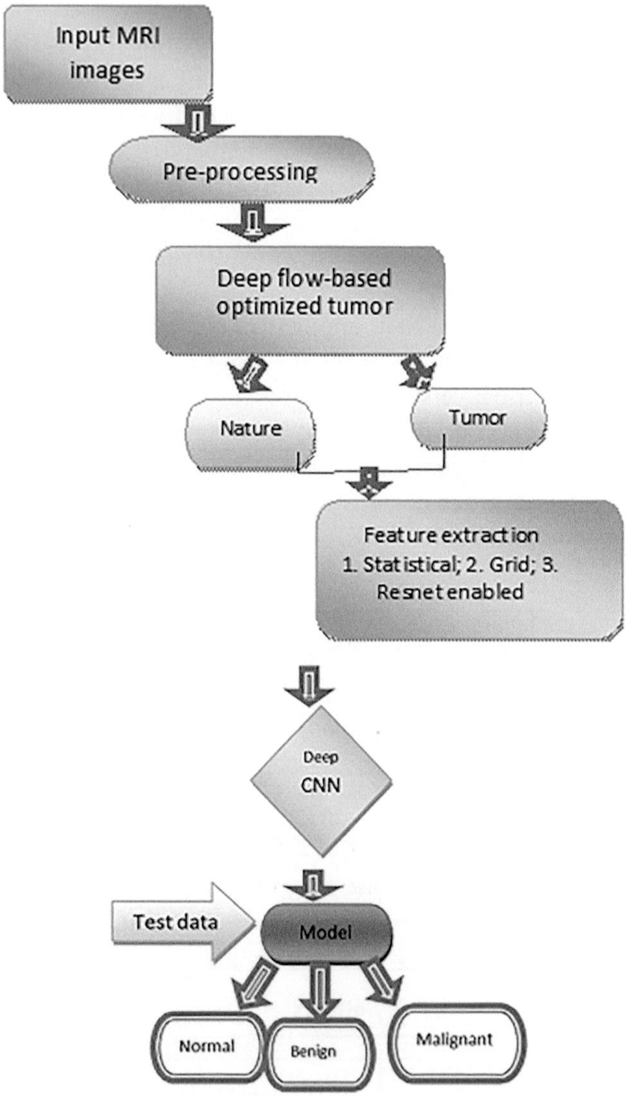

FIGURE 3.3 The suggested deep CNN model's operational flow.

3.4 EXPERIMENTAL RESULTS

3.4.1 INVESTIGATIONAL DATA SET

We used the BraTS dataset for 2020 as part of our test. A total of 2800 images containing T1, T2, and FLAIR tumor types were acquired. This dataset is divided into two classes: 0-class contains images without tumors, while 1-class contains images with tumors. Several tumor datasets and a non-tumor dataset from our input MRI picture are shown in Figures 3.4 and 3.5, respectively.

3.4.2 RESULT AND DISCUSSION

Tables 3.1 and 3.2 present our experimental results with different models, activation functions, and CNN optimizers. AdaMax, an extension of the Adaptive Moment Estimation (Adam) optimizer, was the first approach we tried. We found an accuracy of 98.73% using softmax in the final layer with the optimizer as RMSprop, which was obtained using 2400 training photos, 270 test images, and a split ratio of 9:1. More broadly, it is an extension of the gradient descent optimization technique. In order to avoid overfitting, we removed a small portion of the 2892 total photos (roughly 5% in this case). Figure 3.6 illustrates the outcome of the recommended strategy using 11 epochs. Figure 3.7 displays the non-tumor images.

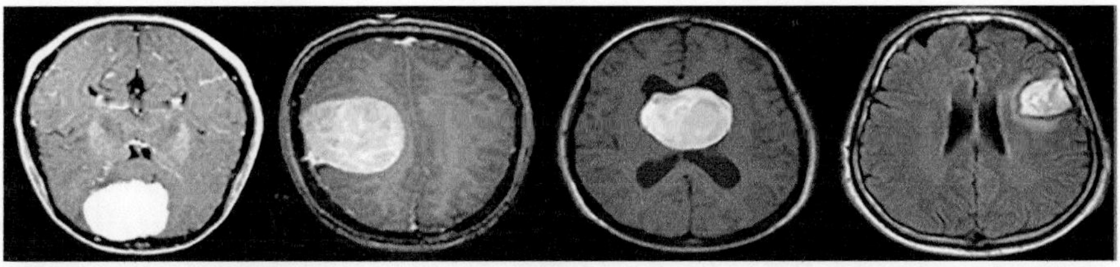

FIGURE 3.4 Tumor images.

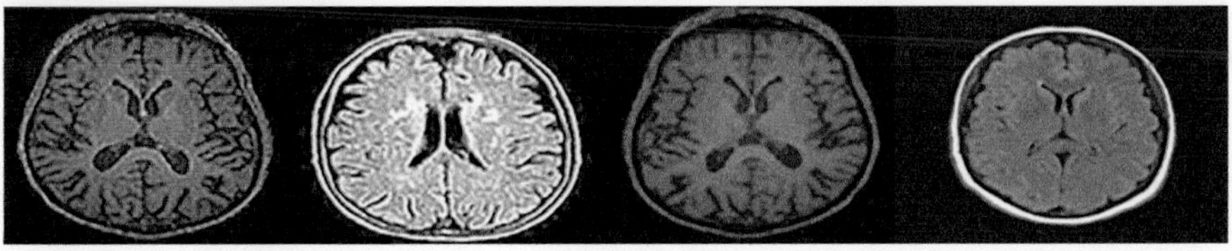

FIGURE 3.5 Non-tumor images.

TABLE 3.1 Contrasting Models

Method of Activating the Final Layer	Optimization Technique Used	Accuracy (%)	Evaluating Accuracy (%)	Evaluation of the Model (%)
Support vector machine	N/A	14.17	19.81	23.16
Sigmoid	Rmsprop	96.62	57.34	67.71
Softmax	AdaMax	97.11	74.01	81.39
Softmax	RMSprop	98.73	93.78	97.70

Note: Some preliminary works of a similar kind can be found from Bhuyan and Borah (2014); Kumar et al. (2014).

TABLE 3.2 Proposed Deep CNN Model's Effectiveness

SL. No.	Trained Data	Test Image	Ratio	Accuracy (%)
1.	2188	541	8:1	98.72
2.	2475	271	9:1	98.73

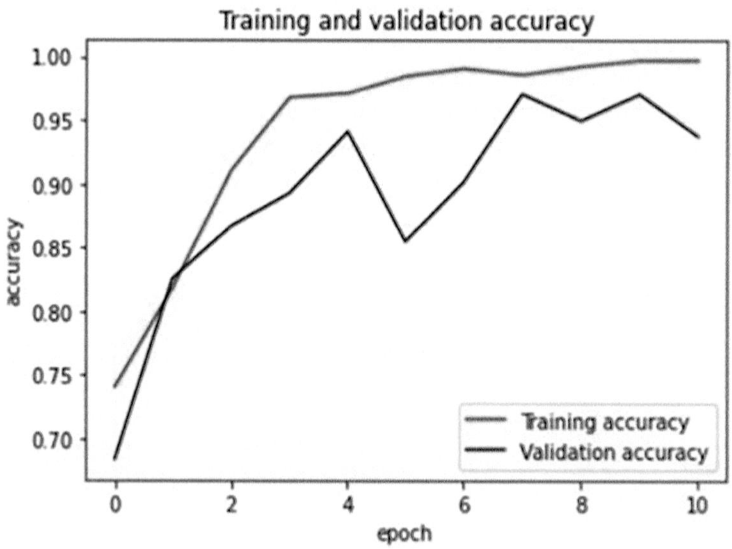

FIGURE 3.6 Training and validation accuracy.

In our model, we obtained 98.73% accuracy, which is higher than the latest results obtained by Seetha et al. (2018) and Hossain et al. (2019), as shown in Table 3.3.

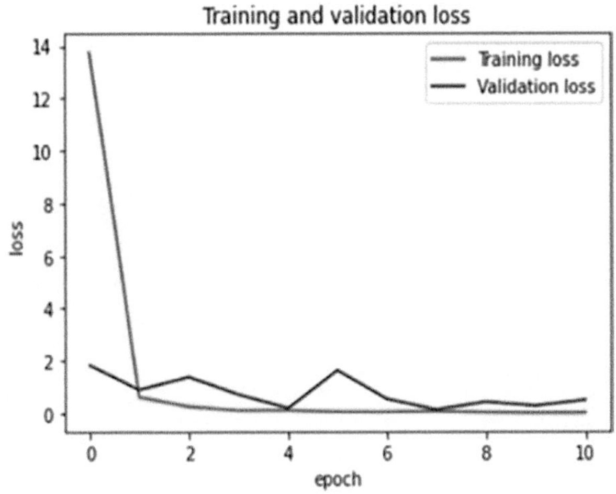

FIGURE 3.7 Loss of training and validation.

TABLE 3.3 Comparison of Performance

Methodology	Accuracy (%)
Seetha and Selvakumar (2018)	97.49
Hosain et al. (2019)	97.86
Our model	98.73

3.5 CONCLUSION

Tumor classification and segmentation are the most typical uses of MRI. Even though CNNs, also known as CNNs, may automatically learn representative complex characteristics for brain regions afflicted by malignancies and healthy tissues straight from multimodal MRI data, we chose to improve the accuracy of CNNs. At the start, we tried to utilize SVM on DeepCNN; however, we got only 20% accuracy. After that, we tried out various settings. We altered the last layer to softmax, and AdaMax was used as the optimizer. After that, we achieved a 98.10% accuracy rate. However, we still wanted more, so we switched to RMSprop for the optimizer, and in the end, we were successful in achieving an output accuracy of 98.73%. Utilizing a 9:1 ratio of 270 frames for testing and 2400 frames for training, we eventually arrived at our result using an 11-epoch strategy. Our model is a deep CNN with nine layers and 14 stages.

DECLARATION

No AI tools or any other similar tools have been used in content creation of this chapter.

KEYWORDS

- **AdaMax**
- **artificial neural network**
- **brain tumor**
- **convolutional neural network**
- **deep learning**
- **gradient algorithm**
- **medical imaging**
- **softmax**
- **support vector machine**

REFERENCES

Bhuyan, R., & Borah, S. (2014). Image segmentation: Computational approaches for medical images. *International Journal of Computer Applications (IJCA), Special Issue on Bioinformatics and Computational Biology,* (3), 13–17. https://www.ijcaonline.org/proceedings/icbcb/number3/17629-1503 (accessed on 11th July 2025).

Cancer.Net Editorial Board. (2021). Brain tumor: Statistics. https://www.cancer.net/cancer-types/brain-tumor/statistics (accessed on 11th July 2025).

Foo, J. L. (2006). A survey of user interaction and automation in medical image segmentation methods (Tech. Rep. ISUHCI20062). Human Computer Interaction Department, Iowa State University.

Gordillo, N., Montseny, E., & Sobrevilla, P. (2013). State of the art survey on MRI brain tumor segmentation. *Magnetic Resonance Imaging, 31*(8), 1426–1438.

Hamamci, A. (2012). Tumor-Cut: Segmentation of brain tumors on contrast enhanced MR images for radiosurgery applications. *IEEE Transactions on Medical Imaging, 31*(3), 790–804.

Havaei, M., Larochelle, H., Poulin, P., & Jadoin, P. M. (2016). Within-brain classification for brain tumor segmentation. *International Journal of Computer Assisted Radiology and Surgery, 11*, 777–788.

Hossain, T., Shishir, F. S., Ashraf, M., Al Nasim, M. D. A., & Shah, F. M. (2019). Brain tumor detection using convolutional neural network. In *1st International Conference on Advances in Science, Engineering and Robotics Technology (ICASERT)*.

Ioffe, S., & Szegedy, C. (2015). Batch normalization: Accelerating deep network training by reducing internal covariate shift. *arXiv preprint*, arXiv:1502.03167. https://arxiv.org/abs/1502.03167 (accessed on 11th July 2025).

Kasban, H., Elbendary, M., & Salama, D. (2015). A comparative study of medical imaging techniques. *International Journal of Information Science and Intelligent Systems, 4*, 37–58.

Kumar, I., Jha, R. S., Kumar, S., & Borah, S. (2014). Brain abnormality detection from MR images using matrix symmetry method. *International Journal of Computer Applications (IJCA) Special Issue on Bioinformatics and Computational Biology,* (3), 22–25. https://www.ijcaonline.org/proceedings/icbcb/number3/17631-1505 (accessed on 11th July 2025).

Maxwell, J. C. (1892). *A Treatise on Electricity and Magnetism* (Vol. 2, 3rd ed., pp. 68–73). Clarendon.

Mustaqeem, A., Javed, A., & Fatima, T. (2012). *International Journal of Image, Graphics and Signal Processing, 10*, 34–39.

Oelze, M. L., Zachary, J. F., & O'Brien, W. D. Jr. (2003). Differentiation of tumor types *in vivo* by scatterer property estimates and parametric images using ultrasound backscatter. *IEEE*, Vol. *1*, 1014–1017.

Prastawa, M., Bullitt, E., & Gerig, G. (2009). Simulation of brain tumors in MR images for evaluation of segmentation efficacy. *Medical Image Analysis, 13*(2), 297–311.

Seetha, J., & Selvakumar Raja, S. (2018). Brain tumor classification using convolutional neural networks. *Biomedical and Pharmacology Journal, 11*, 1457–1461. https://doi.org/10.13005/bpj/1511 (accessed on 11th July 2025).

Surya Prabha, D., & Satheesh Kumar, J. (2016). Performance evaluation of image segmentation using objective methods. *Indian Journal of Science and Technology, 9*(8).

Tolias, G., Sicre, R., & Jégou, H. (2015). Particular object retrieval with integral max-pooling of CNN activations. *arXiv preprint*, arXiv:1511.05879. https://learnopencv.com/batch-normalization-in-deep-networks/ (accessed on 11th July 2025).

Wang, Y. (2020). Modality-pairing learning for brain tumor segmentation. *arXiv preprint*, arXiv:2010.09277v2. https://arxiv.org/abs/2010.09277v2 (accessed on 11th July 2025).

White, D., Houston, A., Sampson, W., & Wilkins, G. (1999). Intra- and inter-operator variations in region-of-interest drawing and their effect on the measurement of glomerular filtration rates. *Clinical Nuclear Medicine, 24*, 177–181.

Zhang, W., Wu, Y., Yang, B., Hu, S., Wu, L., & Dhelim, S. (2021). Overview of multi-modal brain tumor MR image segmentation. *Healthcare, 9*. https://doi.org/10.3390/healthcare9030304 (accessed on 11th July 2025).

CHAPTER 4

Heart Disease Prediction Using Machine Learning

SAGAR JANOKAR,[1] DNYANESHWAR KANADE,[1] MILIND RANE,[1] TEJAS KULKARNI,[2*] VARAD KULKARNI,[3] and YASH KULKARNI[2]

[1]*Department of Electronics and Telecommunication Engineering, BRACT's Vishwakarma Institute of Technology, Pune, Maharashtra, India*

[2]*Department of Information Technology, BRACT's Vishwakarma Institute of Technology, Pune, Maharashtra, India*

[3]*Department of Computer Engineering, BRACT's Vishwakarma Institute of Technology, Pune, Maharashtra, India*

*Corresponding author

ABSTRACT

Heart diseases cause over 17 million deaths annually worldwide, and this number is projected to exceed 70 million in the coming years (Carroll and Miller, 2013). Multiple risk factors such as diabetes, high blood pressure, high cholesterol, and abnormal pulse rates contribute to the complexity of diagnosis (Mohan, Thirumalai, and Srivastava, 2019). Early detection plays a crucial role in preventing fatalities and enabling informed medical decisions (Nikam et al., 2020; Singh and Kumar, 2020). Machine learning (ML) models have emerged as effective tools to support early detection and improve diagnostic accuracy. These models enhance decision-making in the medical field by reducing diagnostic time, increasing efficiency, and lowering costs. This study focuses on improving the accuracy of predictive models for heart disease detection. The proposed model evaluates the likelihood of heart disease based on input factors and predicts the outcome

Intelligent Computing and Analytics in Image Processing. Samarjeet Borah, Jambi Ratna Raja Kumar, Laxmi Bewoor, Rahul Raghvendra Joshi, and Mahendra Prabhakar Deore (Eds.)

DOI: 10.1201/9781779640772-4

in binary terms (Yes/No) (Gavhane et al., 2018). The model achieved a high prediction accuracy of 99.15%. Details of the methods and techniques used in developing this model are discussed in the methodology section.

4.1 INTRODUCTION

The increase in the number of heart diseases and the failure to detect them at an early stage have led to the worsening of heart conditions (Jindal et al. 2021). By processing heart-related data, we can aid in the early detection of heart-related anomalies and thus help in the long-term saving of human lives and better understanding of the condition (Mohan, Thirumalai, and Srivastava, 2019). As the number of deaths has increased in recent years, accurate methods for heart disease prediction are necessary. Our machine learning (ML) model provides a solution to this necessity. Our prediction model, built using the "Random Forest" ML algorithm, predicts based on the inputs given by the user. The working of the Random Forest algorithm is as follows: it builds a large number of decision trees (DTs), and each tree gives a prediction. The value that is predicted more frequently is given as the final prediction. The number of trees for prediction must be predefined in this algorithm (Garg et al. 2021). We have used 500 trees to give the prediction. A greater number of trees helps in giving optimal results but makes the model slower. Therefore, we have chosen a median value that not only makes the model accurate but also quicker. "Heart Disease Prediction Using Machine Learning" is a topic widely researched all over the world (Pouriyeh et al., 2017). Since there is a fair amount of research work done on this concept, we carried out a literature survey by reading some of the research papers. We concluded the following from the survey:

- We noticed that different values of the test size of the dataset gave different accuracies.
- In most of the models built, we inferred that the Random Forest algorithm gave the highest accuracy.

The prediction model is built using the Random Forest ML algorithm, and its testing accuracy is 99.15%. A large database is used, containing the health parameters of over 1000 people, which contributes to the high prediction accuracy of our model. Initially, we used a logistic regression (LR) model for prediction, but it was less accurate than Random Forest. A key finding is that the size of the testing data plays a significant role in determining the model's accuracy—the accuracy changes considerably with even a small change in the size of the testing data. By using a large dataset and adjusting the test size, we can achieve the best possible prediction accuracy. By following the above steps and finding a middle ground, we increased our accuracy from 96.69% to 99.15%. Our model also displays some specialist cardiac hospitals/clinics in major cities if it predicts that the user has a heart disease. This can be helpful to users.

4.2 METHODOLOGY

In our proposed method, heart disease can be predicted more accurately and instantly. In order to build this system, we referred to a number of existing projects and learned the prerequisites. The model used preprocessed data to train and test using the Random Forest algorithm. The data was divided into training data (to train the model) and testing data. The testing phase has a significant role in determining the outcome of a predictive model (Rahman et al., 2022). The training data constitutes 65.5% of the entire dataset, while the testing data constitutes 34.5%. This proportion of data splitting gave us the best possible prediction accuracy. The flow of the code is given in Figure 4.1.

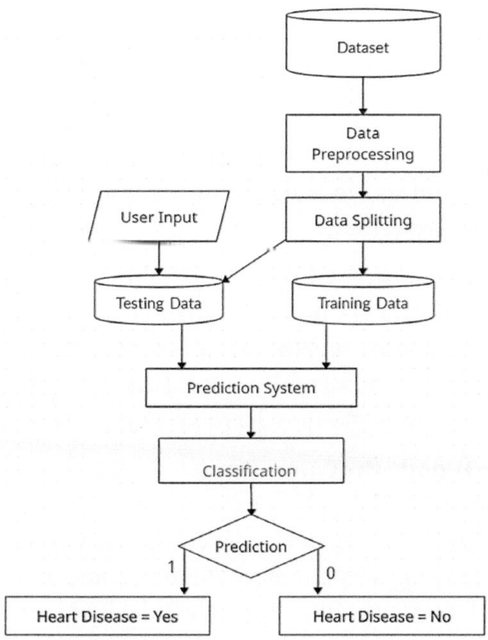

FIGURE 4.1 Flowchart of the method.

The proposed system used the Random Forest algorithm to train the model. Using an IDE (VS Code) and the Python programming language, we trained and predicted the result of a person based on some health-related parameters (Krishnan and Geetha, 2019).

4.2.1 REQUIRED PARAMETERS

1. **Age (in years).**
2. **Gender.**
3. **Chest Pain:**
   i. **Typical Angina:** This is a chest pain type caused by decreased blood flow to the heart. It is often associated with pressure, heaviness, tightness, or a burning sensation in the chest and can sometimes be felt in the neck, jaw, arms, shoulders, or back.
   ii. **Atypical Angina:** This refers to discomfort that doesn't fit the general pattern of typical angina. It generally has different characteristics, such as a stabbing or sharp pain, or it may be felt in unusual locations such as the abdomen, throat, or jaw.
   iii. **Non-Anginal Pain:** This is chest discomfort not caused by reduced blood flow to the heart. It may have other causes such as acid reflux, muscle strain, lung problems, or anxiety, and may be described as a burning, stabbing, or sharp pain.
   iv. **Asymptomatic Chest Pain:** This refers to discomfort in the chest that is not accompanied by other symptoms such as shortness of breath, dizziness, nausea, or sweating. It may be detected incidentally during routine medical checkups or imaging tests and may have various causes.
4. **Resting Blood Pressure (mmHg):** Resting BP refers to the measurement of BP when a person is in a fully relaxed, seated position and not physically exerting themselves. It is typically measured using a sphygmomanometer and is reported in millimeters of mercury (mmHg). The ideal values for systolic and diastolic pressure are 120 mmHg and 80 mmHg, respectively.

5. **Cholesterol (mg/dl):** This is a type of fat necessary for the normal functioning of the body. It is found in various food products such as meat, dairy, and eggs, and is also produced by the liver.
6. **Fasting Blood Sugar:** This refers to the glucose level present in the bloodstream after an individual has abstained from food and drink (other than water) for a minimum of 8 hours.
7. **Electrocardiography Result (Resting):**
   i. **Abnormality in ST-T Wave:** Such abnormalities refer to changes in the electrical activity of the heart that are visible on an electrocardiogram (ECG) as deviations from normal ST segment and T wave patterns. These changes may indicate various cardiac conditions such as injury, inflammation, or electrolyte imbalances.
   ii. **Possible or Definite Left Ventricular Hypertrophy:** This is the thickening of the walls of the lower left heart chamber.
8. **Maximum Heart Rate (Max HR) Achieved:** This is the number of beats the heart makes in 60 seconds under maximum stress.
9. **Exercise-Induced Angina**: This is chest pain or discomfort that occurs during physical activity. It is usually caused by decreased blood flow to the heart due to narrowed or blocked arteries and generally subsides with rest or medication (Ali et al. 2019).
10. **Low ST Segment.**
11. **Slope of ST Segment**: The ST segment has a peak, and its slope is the required parameter.
12. **Colored Vessels in Fluoroscopy:** This refers to the number of colored vessels in a fluoroscopy test. In general, fluoroscopy may be used to visualize the placement of catheters in the coronary arteries, and the tally of vessels colored by dye depends on the extent and location of the arterial blockages being treated. Typically, there are three major coronary arteries that supply blood to the heart.
13. **Thallium Stress Result:** A thallium stress test is a type of nuclear imaging test used to evaluate blood flow to the heart. It involves

injecting a radioactive tracer (thallium) into the bloodstream and then taking images of the heart at rest and during exercise or stress.

14. **Target (Predicted Attribute).**

4.2.2 PROPOSED SYSTEM

The aim of the proposed system is to predict heart diseases more accurately.

4.2.2.1 Random forest algorithm

This is a method used to classify data and train a ML model to make predictions. It is widely used for tasks such as classification and regression (Katarya and Srinivas, 2020). The model has a lower risk of overfitting and runs more efficiently compared to other models. The algorithm works by randomly selecting a subset of features/parameters from the dataset and constructing multiple DTs based on these features (Uddin, Khan, and Hossain, 2019). Each tree is trained using a different subdivision of the data and a random subset of features. The DTs are then combined by aggregating their predictions to produce the final output. The aggregation method can be either a majority vote (for classification) or a mean (for regression). Random Forest is a useful, reliable, and ideal algorithm for handling high-dimensional data in which features are complexly interwoven. One of its advantages is that it can handle missing values and meaningless or corrupted data without overfitting. The algorithm also uses sampling to handle imbalanced datasets. It should be validated on independent datasets to avoid overfitting and ensure generalization (Figures 4.2 and 4.3).

	age	sex	cp	trestbps	chol	fbs	restecg	thalach	exang	oldpeak	slope	ca	thal	target
0	52	1	0	125	212	0	1	168	0	1.0	2	2	3	0
1	53	1	0	140	203	1	0	155	1	3.1	0	0	3	0
2	70	1	0	145	174	0	1	125	1	2.6	0	0	3	0
3	61	1	0	148	203	0	1	161	0	0.0	2	1	3	0
4	62	0	0	138	294	1	1	106	0	1.9	1	3	2	0

FIGURE 4.2 Sample values from dataset.

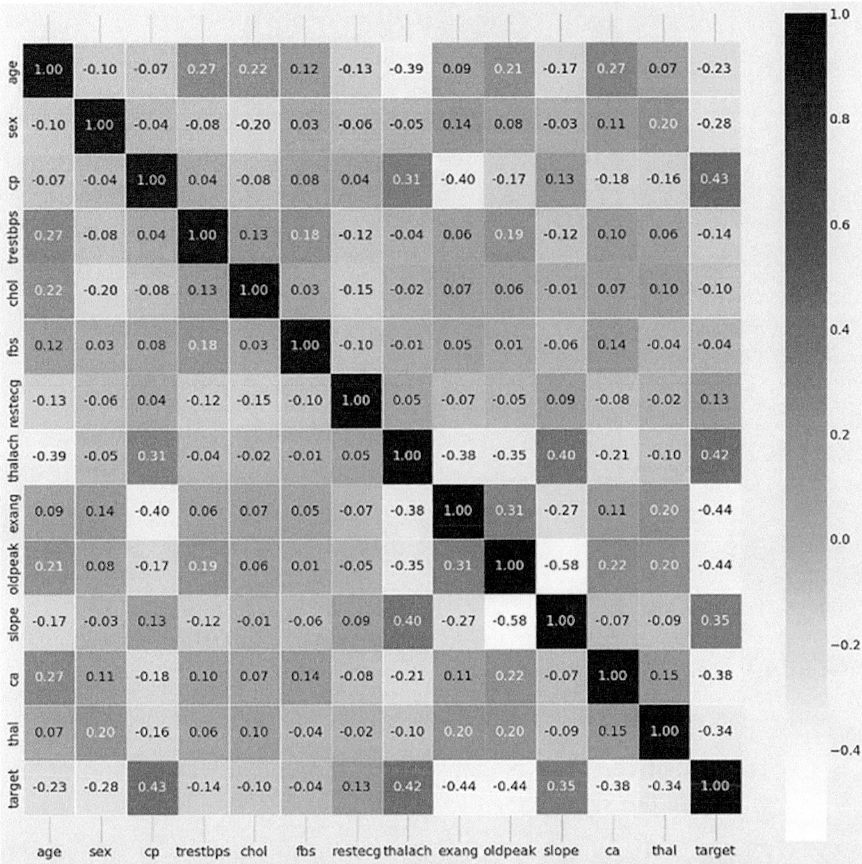

FIGURE 4.3 Correlation matrix.

The correlation matrix obtained from the dataset is shown above. It summarizes the dataset and helps identify as well as visualize patterns in the given data.

4.3 RESULTS AND DISCUSSION

The following are the outcomes that our system produced (Figures 4.4 and 4.5):

- Testing accuracy attained for prediction is 99.15%;
- AUC (area under ROC curve) score of 1.0 is the desired result.

```
age            int64
sex            int64
cp             int64
trestbps       int64
chol           int64
fbs            int64
restecg        int64
thalach        int64
exang          int64
oldpeak      float64
slope          int64
ca             int64
thal           int64
target         int64
dtype: object
(671, 13) (354, 13)
```

FIGURE 4.4 Parameters with data types.

- True Positive, True Negative – The predicted value and actual value are the same

  Example: Predicted = 1, actual = 1; or predicted = 0, actual = 0 (Haq et al., 2019).

- False Positive, False Negative – The predicted value and actual value are dissimilar.

  Example: Predicted = 1, actual = 0; or predicted = 0, actual = 1.

- TP – 177
- FP – 0
- FN – 3
- TN – 174
- Out of 354 test cases, our model gave the correct prediction for 351 cases and incorrect for the remaining, thus giving an accuracy of 99.15%.

$$Precision = \frac{TP}{(TP + FP} = \frac{177}{(177{+}0)} = 1$$

$$Sensitivity = \frac{TP}{(TP + FN)} = \frac{177}{(177{+}3)} = 0.9833$$

$$Accuracy = \frac{(TN + TP)}{(TN + TP + FN + FP)} = \frac{(177{+}174)}{(177{+}174{+}3{+}0)} = 0.9915$$

- A perfect ROC curve is attained.

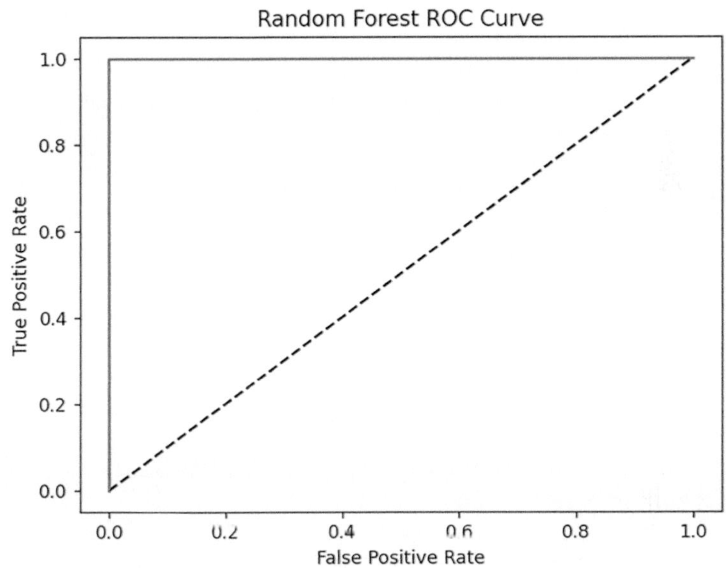

FIGURE 4.5 ROC curve of the predictive model.

- The test size of the data is crucial in establishing the model's correctness.
- In comparison to the accuracy of 85% provided by the logistic regression algorithm, the Random Forest approach provided the best accuracy.

The parameters mentioned in the methodology are defined in the following manner:

1. Age.

2. Sex: 0 for male, 1 for female.
3. Chest pain type:

- 3 = Asymptomatic;
  - 2 = Non-anginal pain;
  - 1 = Atypical angina;
  - 0 = Typical angina.

4. Blood pressure (resting): Measured value.
5. Cholesterol: Measured value.
6. Blood sugar (fasting): 1 for > 120; 0 for < 120.
7. Resting ECG result:

   - 0: Not applicable;
   - 1: Abnormality in ST-T wave;
   - 2: Possible thickening of the walls of the lower left heart chamber.

8. Heart rate (highest): Measured value.
9. Angina induced by exercise: 1 for yes, 0 for no.
10. Low ST segment: 0 for No, 1 for Yes.
11. ST segment slope:

    - 0: Upslope;
    - 1: Flat-sloped;
    - 2: Downslope.

12. Vessels colored by fluoroscopy procedure (0–3).
13. Thallium stress result:

    - 1: Normal;
    - 2: Fixed defect;
    - 3: Reversible defect.

14. Target – predicted attribute – Has a heart disease or not (1 = Yes, 0 = No)

An ideal AUC value is obtained due to a perfect ROC curve.

➢ **Input Required from User:** If the user is predicted to have a heart disease, the system will show some cardiac specialist clinics/hospitals in major cities of India, as shown in Figures 4.6 and 4.7.

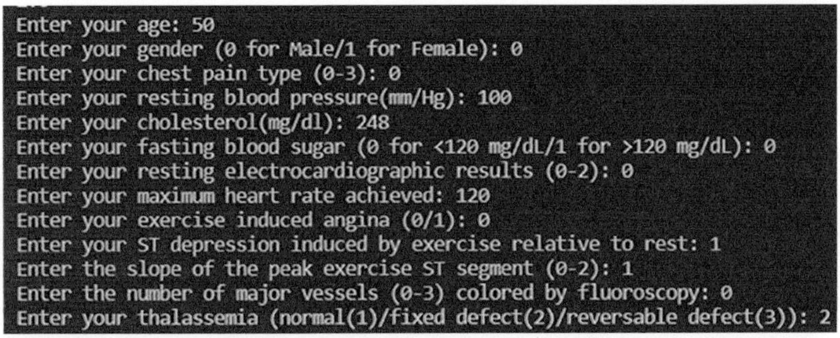

```
Enter your age: 50
Enter your gender (0 for Male/1 for Female): 0
Enter your chest pain type (0-3): 0
Enter your resting blood pressure(mm/Hg): 100
Enter your cholesterol(mg/dl): 248
Enter your fasting blood sugar (0 for <120 mg/dL/1 for >120 mg/dL): 0
Enter your resting electrocardiographic results (0-2): 0
Enter your maximum heart rate achieved: 120
Enter your exercise induced angina (0/1): 0
Enter your ST depression induced by exercise relative to rest: 1
Enter the slope of the peak exercise ST segment (0-2): 1
Enter the number of major vessels (0-3) colored by fluoroscopy: 0
Enter your thalassemia (normal(1)/fixed defect(2)/reversable defect(3)): 2
```

FIGURE 4.6 UI of our predictive model.

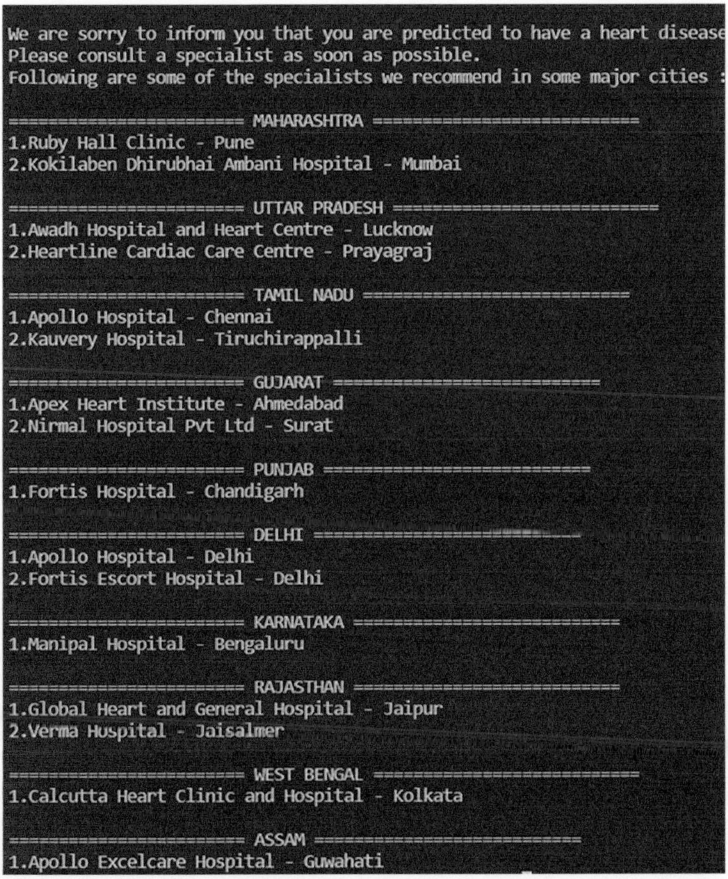

```
We are sorry to inform you that you are predicted to have a heart disease
Please consult a specialist as soon as possible.
Following are some of the specialists we recommend in some major cities :

======================= MAHARASHTRA ===========================
1.Ruby Hall Clinic - Pune
2.Kokilaben Dhirubhai Ambani Hospital - Mumbai

======================= UTTAR PRADESH ===========================
1.Awadh Hospital and Heart Centre - Lucknow
2.Heartline Cardiac Care Centre - Prayagraj

======================= TAMIL NADU ===========================
1.Apollo Hospital - Chennai
2.Kauvery Hospital - Tiruchirappalli

======================= GUJARAT ===========================
1.Apex Heart Institute - Ahmedabad
2.Nirmal Hospital Pvt Ltd - Surat

======================= PUNJAB ===========================
1.Fortis Hospital - Chandigarh

======================= DELHI ===========================
1.Apollo Hospital - Delhi
2.Fortis Escort Hospital - Delhi

======================= KARNATAKA ===========================
1.Manipal Hospital - Bengaluru

======================= RAJASTHAN ===========================
1.Global Heart and General Hospital - Jaipur
2.Verma Hospital - Jaisalmer

======================= WEST BENGAL ===========================
1.Calcutta Heart Clinic and Hospital - Kolkata

======================= ASSAM ===========================
1.Apollo Excelcare Hospital - Guwahati
```

FIGURE 4.7 List of hospitals/clinics displayed.

4.4 CONCLUSION

The goal of this project was to make a system capable of accurately predicting heart disease. The literature review on this topic helped us improve our model. An important inference from the study is that the selection of the test size of data is of paramount significance. The Random Forest classifier gave us the highest accuracy of 99.15%. Out of 354 values, our model gave a wrong prediction for only 3 values. We hope that this project will be helpful for a lot of people.

4.5 FUTURE SCOPE

The aim of this is to aid the maximum number of people in the early detection of heart diseases. We are planning to integrate a location-based system that will also recommend cardiac specialist hospitals around the user. We are also hoping to develop a better user interface by creating a webpage that will be accessible to all. As some parameters may not be available to users, we are planning to modify our model so that it can give a prediction even without some parameters while maintaining the same accuracy.

DECLARATION

No AI tools or any other similar tools have been used in the content creation of this chapter.

KEYWORDS

- **binary prediction**
- **decision trees**
- **heart disease prediction**
- **logistic regression**
- **machine learning**
- **medical decision making**
- **predictive models**
- **random forest algorithm**

REFERENCES

Ali, L., et al. (2019). An optimized stacked support vector machines based expert system for the effective prediction of heart failure. *IEEE Access, 7*, 54007–54014. https://doi.org/10.1109/ACCESS.2019.2909969 (accessed on 11th July 2025).

Bhuyan, R., & Borah, S. (2014). Image segmentation: Computational approaches for medical images. *International Journal of Computer Applications (IJCA), Special Issue on Bioinformatics and Computational Biology,* (3), 13–17. https://www.ijcaonline.org/proceedings/icbcb/number3/17629-1503 (accessed on 11th July 2025).

Carroll, W., & Miller, G. E. (2013). Heart disease among elderly Americans: Estimates for the U.S. civilian noninstitutionalized population, 2010. In *Statistical Brief (Medical Expenditure Panel Survey (US))* [Internet]. Rockville (MD): Agency for Healthcare Research and Quality (US); 2001–. *Statistical Brief #409*. PMID: 29638303.

Garg, A., et al. (2021). *IOP Conference Series: Materials Science and Engineering, 1022*, 012046. https://doi.org/10.1088/1757-899X/1022/1/012046 (accessed on 11th July 2025).

Gavhane, A., Kokkula, G., Pandya, I., & Devadkar, K. (2018). Prediction of heart disease using machine learning. In *2018 Second International Conference on Electronics, Communication and Aerospace Technology (ICECA)*, Coimbatore, India (pp. 1275–1278). https://doi.org/10.1109/ICECA.2018.8474922 (accessed on 11th July 2025).

Haq, A. U., Li, J., Memon, M. H., Memon, M. H., Khan, J., & Marium, S. M. (2019). Heart disease prediction system using model of machine learning and sequential backward selection algorithm for features selection. In *2019 IEEE 5th International Conference for Convergence in Technology (I2CT)*, Bombay, India (pp. 1–4). https://doi.org/10.1109/I2CT45611.2019.9033683 (accessed on 11th July 2025).

Jindal, H., et al. (2021). *IOP Conference Series: Materials Science and Engineering, 1022*, 012072. https://doi.org/10.1088/1757-899X/1022/1/012072 (accessed on 11th July 2025).

Katarya, R., & Srinivas, P. (2020). Predicting heart disease at early stages using machine learning: A survey. In *2020 International Conference on Electronics and Sustainable Communication Systems (ICESC)*, Coimbatore, India (pp. 302–305). https://doi.org/10.1109/ICESC48915.2020.9155586 (accessed on 11th July 2025).

Krishnan, J. S., & Geetha, S. (2019). Prediction of heart disease using machine learning algorithms. In *2019 1st International Conference on Innovations in Information and Communication Technology (ICIICT)*, Chennai, India (pp. 1–5). https://doi.org/10.1109/ICIICT1.2019.8741465 (accessed on 11th July 2025).

Kumar, I., Jha, R. S., Kumar, S., & Borah, S. (2014). Brain abnormality detection from MR images using matrix symmetry method. *International Journal of Computer Applications (IJCA) Special Issue on Bioinformatics and Computational Biology,* (3), 22–25. https://www.ijcaonline.org/proceedings/icbcb/number3/17631-1505 (accessed on 11th July 2025).

Mohan, S., Thirumalai, C., & Srivastava, G. (2019). Effective heart disease prediction using hybrid machine learning techniques. *IEEE Access, 7*, 81542–81554. https://doi.org/10.1109/ACCESS.2019.2923707 (accessed on 11th July 2025).

Nikam, A., Bhandari, S., Mhaske, A., & Mantri, S. (2020). Cardiovascular disease prediction using machine learning models. In *2020 IEEE Pune Section International Conference (PuneCon)*, Pune, India (pp. 22–27). https://doi.org/10.1109/PuneCon50868.2020.9362367 (accessed on 11th July 2025).

Pouriyeh, S., Vahid, S., Sannino, G., De Pietro, G., Arabnia, H., & Gutierrez, J. (2017). A comprehensive investigation and comparison of machine learning techniques in the domain of heart disease. In *2017 IEEE Symposium on Computers and Communications (ISCC)*, Heraklion (pp. 204–207). https://doi.org/10.1109/ISCC.2017.8024530 (accessed on 11th July 2025).

Rahman, M., Rana, M., Nur-A-Alam, M., Khan, M., & Mohi Uddin, K. M. (2022). A web-based heart disease prediction system using machine learning algorithms. *International Journal of Computer Applications, 12*, 64–80.

Singh, A., & Kumar, R. (2020). Heart disease prediction using machine learning algorithms. In *2020 International Conference on Electrical and Electronics Engineering (ICE3)*, Gorakhpur, India (pp. 452–457). https://doi.org/10.1109/ICE348803.2020.9122958 (accessed on 11th July 2025).

Uddin, S., Khan, A., & Hossain, M. (2019). Comparing different supervised machine learning algorithms for disease prediction. *BMC Medical Informatics and Decision Making, 19*, 281. https://doi.org/10.1186/s12911-019-1004-8 (accessed on 11th July 2025).

CHAPTER 5

Deep Learning-Based Pneumonia Detection System Using VGG16 and Inception V3 for NIH Chest X-Ray Images

SHEELA CHINCHMALATPURE,* CHETANYA RATHI, HARSH DHABEKAR, and HARSHIT MUNDHRA

Department of Artificial Intelligence and Data Science, BRACT's Vishwakarma Institute of Technology, Pune, Maharashtra, India

*Corresponding author

ABSTRACT

Due to its potential for early diagnosis and prompt treatment, medical imaging-based pneumonia detection has attracted considerable attention lately. Our methodology involves domain adaptation to address the domain shift between publicly available chest X-ray data and our specific dataset. We strategically combine these features with deep learning (DL) representations by integrating the scale-steady model. The models are fine-tuned for pneumonia detection within our dataset.

Our quantitative results demonstrate the unique capabilities of each model in classifying chest X-ray images. VGG16 exhibits effectiveness in identifying non-pneumonia cases, while Inception V3 achieves a balanced classification with improved accuracy in detecting pneumonia cases. An ensemble model combining both architectures presents a trade-off between accuracy and recall.

By demonstrating the potential of DL, domain adaptation, and feature fusion for precise pneumonia detection, this research advances the field of medical image analysis.

Intelligent Computing and Analytics in Image Processing. Samarjeet Borah, Jambi Ratna Raja Kumar, Laxmi Bewoor, Rahul Raghvendra Joshi, and Mahendra Prabhakar Deore (Eds.)

DOI: 10.1201/9781779640772-5

5.1 INTRODUCTION

The early and accurate diagnosis of medical conditions is a critical aspect of healthcare, as it enables timely intervention and improved patient outcomes. Pneumonia, a prevalent respiratory disease, is no exception to this paradigm. Chest X-ray imaging offers a noninvasive means to assess lung health. This has revolutionized the field, allowing for the development of sophisticated automated diagnostic systems. In this study, we embark on a research effort to develop a robust Pneumonia Detection System utilizing deep learning (DL) architectures, VGG16 and Inception V3.

The advent of DL has brought forth the capability to model intricate relationships within medical images, potentially outperforming traditional image analysis methods. Our research focuses on the utilization of the VGG16 and Inception V3 architectures, both pre-trained on extensive image datasets, to harness their capacity for feature extraction. These architectures are also fine-tuned specifically for pneumonia detection within our dataset, incorporating domain adaptation techniques to bridge the gap between the source domain (publicly available chest X-ray data) and the target domain (our specific dataset). Addressing the domain shift is pivotal to enhancing the performance of these models in the specific context of pneumonia detection.

To further enhance the feature extraction process, we integrate the scale-invariant feature transform (SIFT) algorithm, a classical computer vision technique known for its robustness to scale and rotation variations. The combination of DL and traditional feature extraction methods is aimed at capturing a comprehensive set of discriminative features, thereby improving the overall diagnostic performance.

This research not only aims to develop an effective pneumonia detection system but also provides insights into the interplay between DL and traditional computer vision techniques in the context of medical image analysis. The findings presented in this paper shed light on the strengths and limitations of the proposed approach, the performance of individual models, and the potential trade-offs in ensemble models, thereby contributing to the ongoing discourse on the application of DL in healthcare. Early and accurate pneumonia detection can significantly impact patient care and outcomes, making this research highly relevant in the field of medical imaging and healthcare technology.

5.2 LITERATURE SURVEY

The rapid evolution of deep learning has revolutionized medical imaging, particularly in the automated detection of pneumonia through chest X-ray analysis. Recent advancements in deep learning have significantly improved pneumonia detection using chest X-ray images. Gupta et al. (2020) compared deep learning models and found that CNN-based architectures outperform traditional methods in terms of accuracy. Kaushik et al. (2020) also highlighted the potential of CNNs, demonstrating their effectiveness in feature extraction and classification. Kundu et al. (2021) proposed an ensemble of deep models, showing improved diagnostic performance. Similarly, Račić et al. (2021) introduced an efficient deep neural network ensemble that enhanced detection accuracy. Okeke et al. (2019) emphasized model efficiency for healthcare applications, proposing a lightweight approach with high precision. Lopes et al. (2020) provided a comprehensive review of deep learning techniques, highlighting various model architectures and challenges. Patel et al. (2021) explored transfer learning strategies, making model training faster and more effective with limited data. Rajasenbagam et al. (2021) integrated CNNs with content-based image retrieval to further refine results. The benchmark CheXNet model by Rajpurkar et al. (2017) demonstrated radiologist-level performance using deep learning. Furthermore, Grad-CAM by Selvaraju et al. (2017) was used for visual explanations, increasing model interpretability. Based on these studies the promise of deep learning can be seen in automating and improving pneumonia diagnosis from chest X-rays.

5.3 PROPOSED METHODOLOGY

This exploration section outlines a comprehensive methodology for detecting pneumonia using DL methods, specifically focusing on the VGG16 and Inception V3 architectures, domain adaptation, and the integration of SIFT.

The process begins with the acquisition of a chest X-ray dataset, carefully chosen to include images and essential patient metadata. This dataset undergoes preprocessing, including resolution standardization, pixel value normalization, metadata processing, and data augmentation for diversity improvement.

For the core methodology, pre-trained DL models VGG16 and Inception V3 are fine-tuned for pneumonia detection within the dataset.

A fusion technique is implemented to combine their predictions at a specific network layer, leveraging their complementary information.

Domain adaptation is a pivotal aspect, addressing differences between the source domain (publicly available chest X-ray data) and the target domain. Domain shift analysis identifies variations in image quality, patient demographics, and more. Domain adaptation techniques, like adversarial training, are applied to align these domain distributions, improving model performance.

SIFT is integrated for feature extraction from chest X-ray images, with careful parameter configuration. These SIFT features are fused with DL features and subjected to any necessary normalization or dimensionality reduction.

The training phase involves optimization algorithms with predefined hyperparameters and specialized loss functions. Model evaluation uses standard criteria, such as accuracy, sensitivity, specificity, ROC curves, and AUC. K-fold cross-validation ensures robustness.

In summary, the methodology offers a holistic approach to pneumonia detection, using DL models, domain adaptation, and SIFT for robust and accurate results. Rigorous preprocessing, model fine-tuning, and evaluation criteria contribute to the study's comprehensiveness and reliability (Figures 5.1 and 5.2).

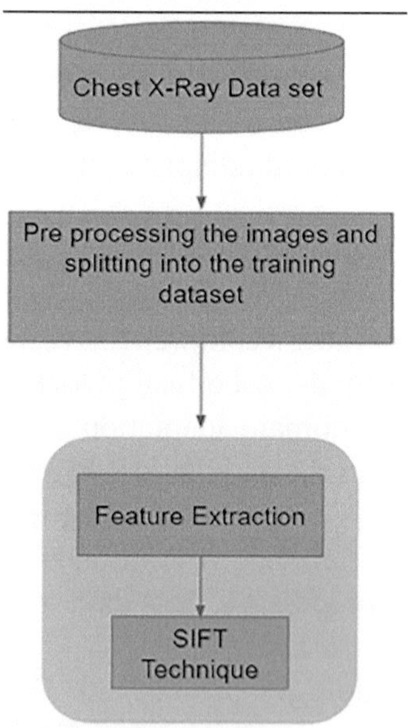

FIGURE 5.1 Feature extraction and preprocessing.

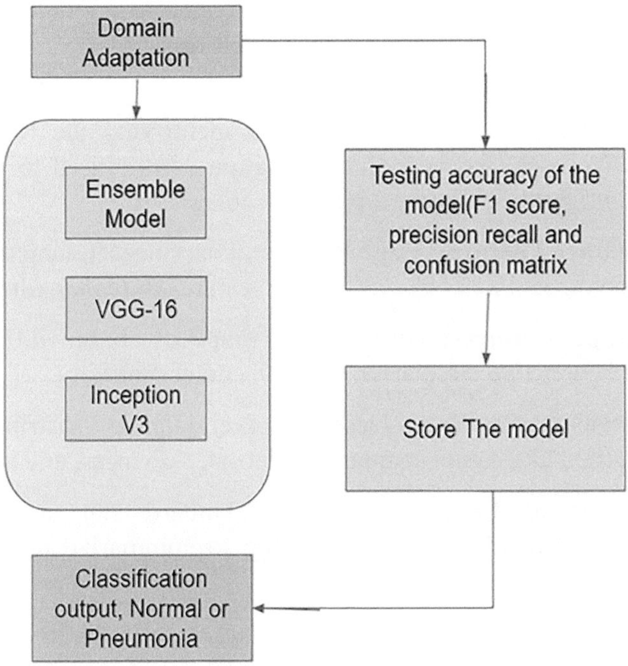

FIGURE 5.2 Model training and testing.

1. **Preprocessing:** Prior to feature extraction, it is imperative to preprocess the images. This could entail tasks such as resizing, normalization, noise reduction, and contrast enhancement to ensure uniformity and quality within the dataset (Figure 5.3).

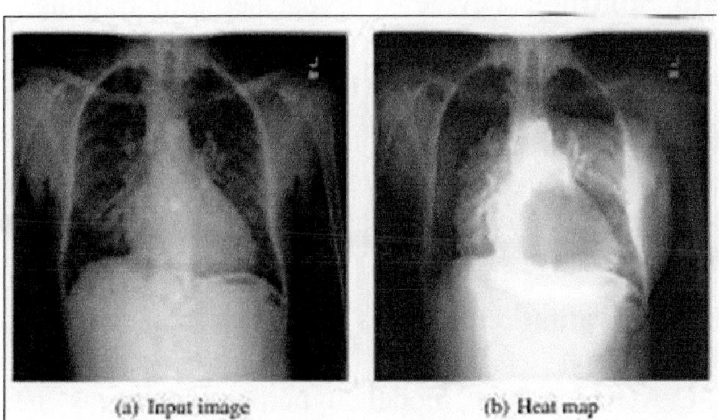

FIGURE 5.3 (a) Input image; and (b) preprocessed image.

2. **Feature Extraction:** Within the context of feature extraction, several categories of features are relevant.
3. **Region of Interest (ROI) Selection:** Pneumonia typically affects specific regions of the lung, so identifying the ROI is vital. Techniques like image segmentation can be used to isolate and highlight the relevant lung area.
4. **Texture Features:** Capture textural variances in lung tissue using techniques like *Haralick* texture features or *Gabor* filters.
5. **Shape Features:** Quantify the shape of abnormalities through attributes like circularity, solidity, and eccentricity.
6. **Intensity Features:** Describe pixel intensity distribution with metrics like mean, standard deviation, skewness, and kurtosis.
7. **Statistical Features:** Include standard statistical measures like mean, median, and variance to summarize pixel intensity distribution.
8. **Frequency Domain Features:** Capture patterns in images using techniques such as Fourier or wavelet transforms.
9. **Dimensionality Reduction:** If the feature space becomes unwieldy due to a large number of extracted features, dimensionality reduction methods like principal component analysis (PCA) can be applied to streamline the dataset while preserving important information.
10. **Data Splitting:** Divide your dataset into training, validation, and testing subsets to facilitate model training and performance evaluation.
11. **Model Building:** Employ DL models to train on the feature vectors created in the previous step. Convolutional neural networks (CNNs) are common choices for detection of pneumonia in medical images.
12. **Model Evaluation:** Assess the performance of models with metrics such as accuracy, recall, precision, F1 score, and ROC-AUC. Adjust model hyperparameters as necessary to improve performance.

13. **Interpretation:** Some models provide insights into which features extracted in the previous step were most influential in making diagnostic predictions, shedding light on the underlying patterns in the data.
14. **Validation:** Ensure the model's robustness and generalizability by validating its performance on new, unseen data. It's important that the success of the feature extraction and the selection of appropriate features have a significant impact on the overall effectiveness of your pneumonia detection model. The choice of features should align with the imaging modality, dataset characteristics, and specific objectives of the application.
15. **SIFT (Scale-Invariant Feature Transform):** This is a foundation of computer vision, known for its robust keypoint detection, feature extraction, and image matching capabilities. It excels at identifying scale- and rotation-invariant keypoints and associating them with descriptors capturing local image structure. In medical image analysis, SIFT is useful for extracting distinctive features from X-rays or CT scans, aiding in image registration, object recognition, and retrieval in medical diagnostics. However, for direct disease prediction, specialized attributes like texture and shape often take priority. Domain adaptation for pneumonia prediction models addresses domain shifts due to demographic and equipment variations across healthcare institutions. Techniques encompass feature-based, patient-based, and model-based approaches, aiming to reduce distribution discrepancy and enhance model adaptability. Evaluation relies on standard criteria, with critical considerations of data privacy and bias in healthcare applications. In summary, domain adaptation is vital for robust and reliable pneumonia prediction models across different clinical environments, demanding both rigorous validation and ethical foresight.
16. **VGG 16:** It is a widely used CNN architecture that plays a vital role in pneumonia disease prediction from medical images. With its 16 weight layers, it excels in the extraction of intricate features from chest X-rays or CT scans, making it a valuable tool for this application. VGG16 automatically learns and extracts relevant features, including texture, shapes, and pixel intensities, which are

vital for detecting pneumonia. Additionally, pre-trained VGG16 models, originally trained on large image datasets, facilitate transfer learning, giving it a head start in understanding general image features and adapting to the specific nuances of medical imaging. In medical datasets, which tend to be relatively small, fine-tuning the pre-trained VGG16 model allows it to specialize in pneumonia-related features efficiently. The deep architecture of VGG16 is known for capturing intricate image details, leading to high predictive accuracy and robustness in pneumonia detection. Evaluation of pneumonia prediction models using VGG16 generally involves criteria such as accuracy, recall, precision, and ROC-AUC, ensuring their reliability in diagnosing the disease. VGG16 simplifies the process of feature extraction and disease prediction, particularly in the medical domain, where its DL capabilities prove invaluable in detecting pneumonia from radiological images with high accuracy and sensitivity.

17. **Inception V3:** Inception V3, a state-of-the-art CNN architecture, is a powerful tool for pneumonia disease prediction from medical images. With its intricate Inception modules, it excels in capturing diverse features from chest X-rays or CT scans. Inception V3's unique architecture incorporates multiple filter sizes within a single layer, allowing it to detect features at different scales simultaneously. Its deep layers are capable of learning intricate patterns within the images, enhancing its ability to identify pneumonia-related anomalies. In medical imaging, Inception V3 has proven effective due to its high accuracy and robustness. Standard measures including accuracy, recall, precision, and ROC-AUC are employed to assess its predictions, guaranteeing a reliable diagnosis of pneumonia.

18. **Ensemble Model:** An ensemble model combining VGG16 and Inception V3, both powerful CNNs, offers a robust approach to pneumonia disease prediction from medical images. VGG16 excels in capturing intricate details and shapes, while Inception V3's unique architecture handles diverse features. The ensemble leverages the strengths of both models, enhancing predictive accuracy by aggregating their outputs. By fusing their complementary abilities, the model not only improves diagnostic reliability but also

effectively addresses different aspects of the disease. The efficiency of the ensemble is ensured by evaluation using measures such as accuracy, recall, precision, and ROC-AUC, making it a helpful aid for accurate and robust pneumonia diagnosis in medical imaging.

5.4 RESULTS AND DISCUSSION

The quantitative results of our DL method for pneumonia diagnosis using the VGG16 and Inception V3 models, together with the ensemble model that combines these two architectures, are shown in this section. Precision, recall, and F1-score are three important performance indicators that provide information on how well the models classify chest X-ray images.

For the VGG16 model, we observed a threshold of 0.59, resulting in a precision of 0.35, recall of 0.51, and an F1-score of 0.41. The confusion matrix for this model reveals its effectiveness in identifying pneumonia cases (pneumonia) with a recall of 0.50, albeit with a relatively lower precision of 0.34. On the other hand, the model's ability to accurately identify non-pneumonia cases (no pneumonia) is demonstrated by a recall of 0.78 and a higher precision of 0.87 (Figures 5.4 and 5.5).

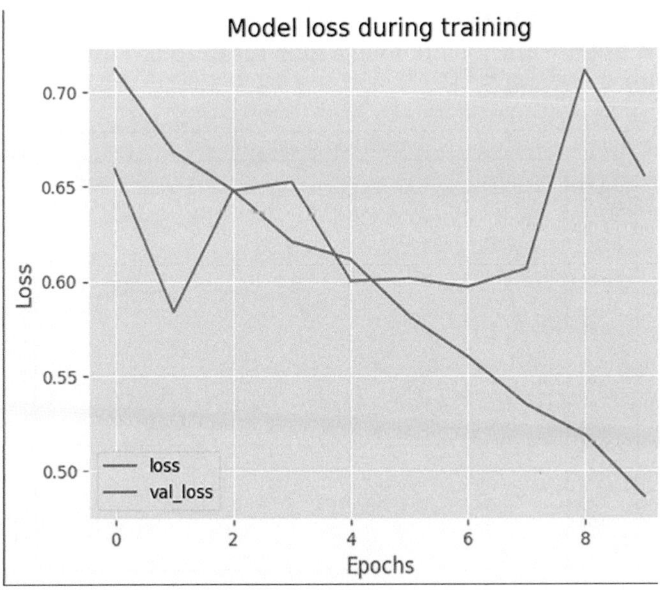

FIGURE 5.4 VGG16 model loss during training.

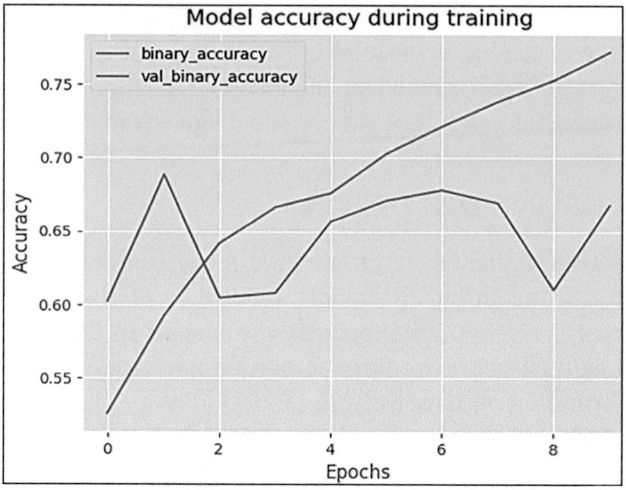

FIGURE 5.5 VGG16 model accuracy during training.

The Inception V3 model, with a threshold set at 0.52, yielded improved performance with a precision of 0.40, recall of 0.44, and an F1-score of 0.42. The confusion matrix illustrates a balanced classification, showcasing its proficiency in distinguishing between pneumonia and non-pneumonia cases. Notably, this model achieved a higher precision (0.40) in detecting pneumonia cases while maintaining a high recall (0.85) for non-pneumonia cases (Figures 5.6 and 5.7).

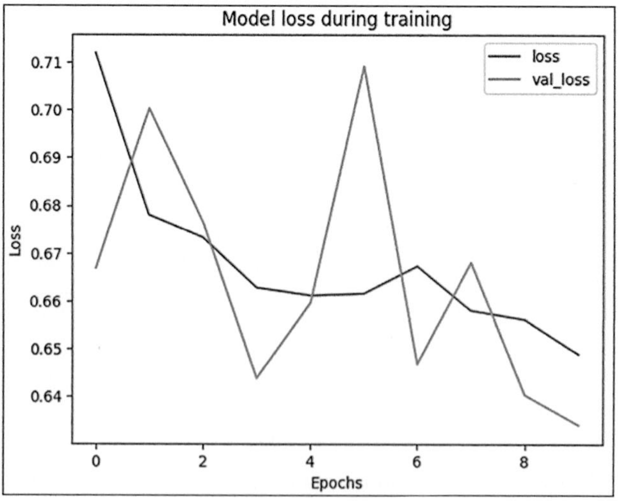

FIGURE 5.6 Inception V3 model loss during training.

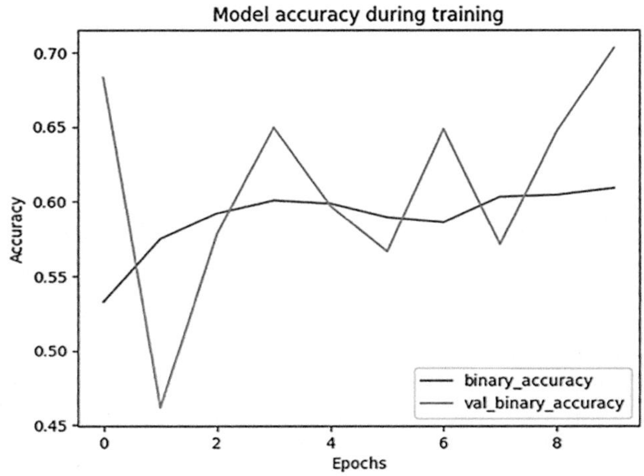

FIGURE 5.7 Inception V3 model accuracy during training.

The ensemble model, with a threshold of 0.54, demonstrated precision, recall, and F1-score values of 0.31, 0.64, and 0.42, respectively. The confusion matrix for the ensemble model indicates a significant improvement in recall for pneumonia cases (0.63); yet, this comes at the cost of a lower precision (0.30). Furthermore, it exhibits a decreased precision (0.31) in identifying non-pneumonia cases but still maintains a moderate recall (0.67) (Figures 5.8 and 5.9).

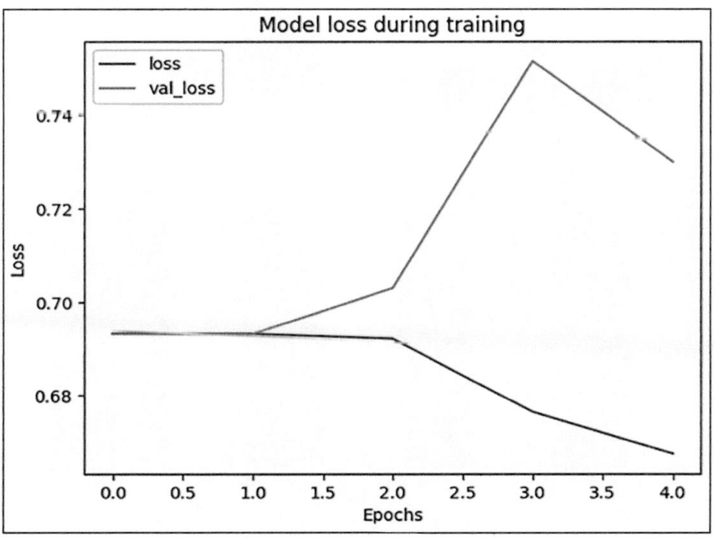

FIGURE 5.8 Ensemble model loss during training.

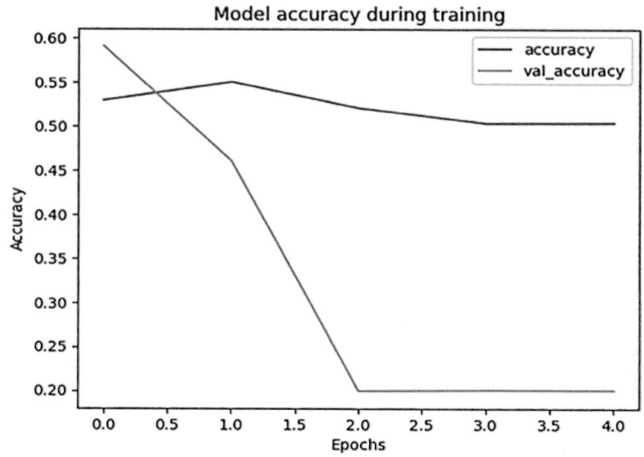

FIGURE 5.9 Ensemble model accuracy during training.

Overall, these quantitative results underline the distinct characteristics of each model in pneumonia detection. While Inception V3 emerges as the superior performer in terms of precision and F1-score, the ensemble model presents a trade-off between precision and recall. These findings are essential for understanding the models' behavior and will guide future efforts aimed at optimizing pneumonia detection systems (Figure 5.10).

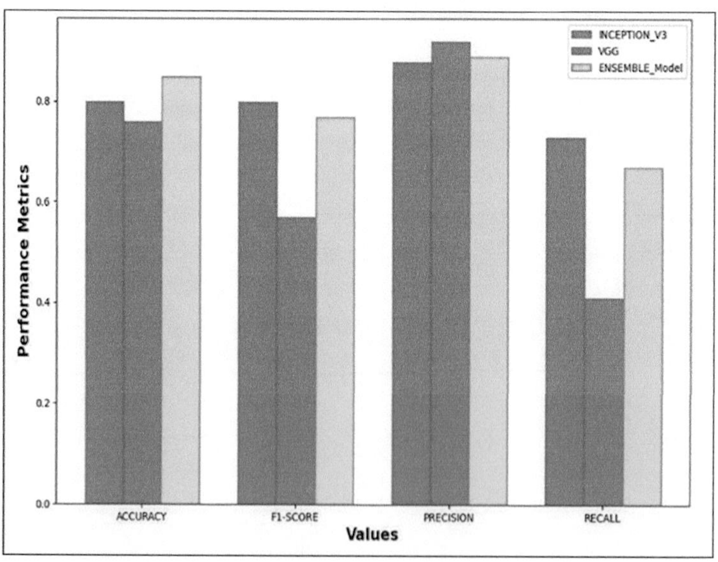

FIGURE 5.10 Comparison.

5.4 CONCLUSION

In summary, our study demonstrates that Inception V3 outperforms VGG16 with a more balanced precision–recall trade-off, resulting in a competitive F1-score. The ensemble model, while enhancing recall, exhibits a significant decline in precision, indicating that it may classify more pneumonia cases but at the expense of increased false positives (FPs). These findings underscore the importance of a detection system when selecting a model. Researchers and practitioners should carefully weigh the trade-offs between precision and recall in their chosen approach, depending on the clinical context and the associated consequences of FPs and false negatives (FNs). Further research is necessary to fine-tune and optimize ensemble models to strike a more equitable balance between these essential metrics, considering the specific goals and requirements of pneumonia detection.

DECLARATION

No AI tools or any other similar tools have been used in content creation of this chapter.

KEYWORDS

- **domain adaptation**
- **ensemble model**
- **inception V3**
- **medical image analysis**
- **pneumonia detection**
- **scale-invariant feature transform**
- **VGG16**

REFERENCES

Gupta, S., Anand, D., & Mittal, A. (2020). Performance comparison of deep learning models for chest X-ray based pneumonia detection. In *2020 6th International Conference on Advanced Computing and Communication Systems (ICACCS)* (pp. 1072–1076). IEEE. https://doi.org/10.1109/ICACCS48705.2020.9074364 (accessed on 11th July 2025).

Kaushik, V. S., Nayyar, A., Kataria, G., & Jain, R. (2020). Pneumonia detection using convolutional neural networks (CNNs). *ResearchGate*. https://www.researchgate.net/publication/340582552 (accessed on 11th July 2025).

Kundu, R., Das, R., Geem, Z. W., Han, G. T., & Sarkar, R. (2021). Pneumonia detection in chest X-ray images using an ensemble of deep learning models. *PLOS ONE, 16*(9), e0256857. https://doi.org/10.1371/journal.pone.0256857 (accessed on 11th July 2025).

Lopes, F. M., de Santana, L. F., de Santana, R. S., et al. (2020). Deep learning in pneumonia detection on chest X-rays: A comprehensive review. *Journal of Healthcare Engineering, 2020*. https://doi.org/10.1155/2020/8890873 (accessed on 11th July 2025).

Okeke, S., Sain, M., Maduh, U. J., & Jeong, D. U. (2019). An efficient deep learning approach to pneumonia classification in healthcare. *Journal of Healthcare Engineering, 2019*, Article ID 4180949. https://doi.org/10.1155/2019/4180949 (accessed on 11th July 2025).

Patel, M., Sojitra, A., Patel, Z., & Bohara, M. H. (2021). Pneumonia detection using transfer learning. *International Journal of Engineering Research & Technology (IJERT), 10*(10). https://www.ijert.org/pneumonia-detection-using-transfer-learning (accessed on 11th July 2025).

Račić, L., Popovic, T., Cakic, S., & Šandi, S., Suryaa, V. S., Annie, A. X. R., & Aiswarya, M. S. (2021). Efficient DNN ensemble for pneumonia detection in chest X-ray images. *International Journal of Advanced Computer Science and Applications (IJACSA), 12*(10). https://doi.org/10.14569/IJACSA.2021.0121020 (accessed on 11th July 2025).

Rajasenbagam, T., Jeyanthi, S., & Arun Pandian, J. (2021). Detection of pneumonia infection in lungs from chest X-ray images using deep convolutional neural network and content-based image retrieval techniques. *Journal of Ambient Intelligence and Humanized Computing*. https://doi.org/10.1007/s12652-021-03064-5 (accessed on 11th July 2025).

Rajpurkar, P., Irvin, J., Zhu, K., et al. (2017). CheXNet: Radiologist-level pneumonia detection on chest X-rays with deep learning. *arXiv preprint*, arXiv:1711.05225. https://arxiv.org/abs/1711.05225 (accessed on 11th July 2025).

Selvaraju, R. R., Cogswell, M., Das, A., Vedantam, R., Parikh, D., & Batra, D. (2017). Grad-CAM: Visual explanations from deep networks via gradient-based localization. In *Proceedings of the IEEE International Conference on Computer Vision (ICCV)* (pp. 618–626). https://doi.org/10.1109/ICCV.2017.74 (accessed on 11th July 2025).

Shin, H. C., Roth, H. R., Gao, M., et al. (2016). Deep convolutional neural networks for computer-aided detection: CNN architectures, dataset characteristics and transfer learning. *IEEE Transactions on Medical Imaging, 35*(5), 1285–1298.

CHAPTER 6

Comparative Study Using Deep Learning for Detecting Plant Type and Leaf Disease

PUNAM KHANDAR,[1,*] SANDEEP NANDA,[2] SARAT CHANDRA SWAIN,[3] and DEEPALI BHENDE[4]

[1]*Department of Computer Applications, Shri Ramdeobaba College of Engineering and Management (RCOEM), Nagpur, Maharashtra, India*

[2]*K. R. Mangalam University, Gurugram, Haryana, India*

[3]*School of Electrical Engineering, KIIT Deemed to be University, Bhubaneswar, Odisha, India*

[4]*Department of Computer Science, Raisoni University, Saikheda, Madhya Pradesh, India*

*Corresponding author

ABSTRACT

In recent years, plant diseases have been increasing significantly due to severe climatic changes and a lack of immunity in plants. Large-scale crop destruction results from this, reducing cultivation and ultimately affecting farmers' income. Identifying and treating these diseases has become difficult due to the rapid emergence of new diseases and limited human understanding. However, various studies suggest that the texture and visual characteristics of leaves are useful for identifying disease types. Hence, the solution to this challenge may lie in combining deep learning (DL) and computer vision. In the present work, we propose a model based on DL.

Intelligent Computing and Analytics in Image Processing. Samarjeet Borah, Jambi Ratna Raja Kumar, Laxmi Bewoor, Rahul Raghvendra Joshi, and Mahendra Prabhakar Deore (Eds.)

DOI: 10.1201/9781779640772-6

It is trained on and oriented toward analyzing public datasets comprising images of both healthy and diseased plant leaves. The model accomplishes this by categorizing images of infected leaves.

6.1 INTRODUCTION

Every day and for every generation, agriculture serves as the cornerstone of a nation's economic growth by providing food, raw materials, and fuel. Approximately 66% of people depend on agriculture, either directly or indirectly. Agriculture is struggling to keep up with demand due to the enormous rise in global population. Food security remains under threat from several factors, including climate change, a decline in pollinators, plant diseases, and inadequate irrigation. Plant diseases negatively affect both the quality and yield of agricultural production. The impact is particularly severe for small-scale farmers, who rely on successful cultivation for their livelihoods, and it also poses a broader threat to global food security. By enabling the monitoring of large fields and the automatic detection of disease symptoms at early stages, such technologies can mitigate damage once symptoms manifest on leaf tissue (Arivazhagan et al., 2013).

The purpose of this chapter is to present an application that predicts crop disease type using leaf textural similarity. The model is trained using a publicly accessible dataset comprising images of both healthy and diseased plant leaves. Early detection of plant diseases can help prevent further crop damage and is beneficial for maintaining sustainable cultivation.

The rest of the chapter is organized into the following segments. Section 2 includes related research in this field. Section 3 covers the proposed model and the approach taken to solve the problem. Section 4 details the experiments conducted and the results obtained. The chapter concludes with a summary of the findings, including a comparison of models and their future applicability.

6.2 LITERATURE SURVEY

Before identifying the class of illness, we must first identify the plant species. In the research, Abdul Kadir employed pixel values of the leaves to create color attributes such as mean, standard deviation, skewness, and kurtosis. The author combined various features obtained from the grey level co-occurrence matrix (GLCM) operations (Kadir, 2014) and then analyzed the image texture. A GLCM is developed by examining combinations of pixels with specific values and their spatial relationships within the image to generate statistical data.

According to Naik and Sivappagari (2016), the disease has been detected using a neural network. In this paper, if the leaf is contaminated, additional processing is carried out to determine the disease. The research proposes a hybrid approach, integrating support vector machine (SVM) with a genetic algorithm to reduce loss and classify disease types. This novel method employs a genetic algorithm to improve the loss function, mirroring the principles of natural selection, where only robust parameters persist. Papers by Filev et al. (2014), *SVM Classifier Based Grape Leaf Disease Detection* (2016), and Zhang and Zhang (2010) also use the SVM classification approach for the identification and categorization of leaf diseases in grapes, tomatoes, cucumbers, and various other plant leaves. Here, the diseased region is first identified by the authors using segmentation through K-means clustering, followed by extraction of texture and color features. Using the above classification technique, the type of plant disease is then determined.

Paper by Olmschenk, Tang, and Zhu (2018), semi-supervised learning is employed for categorizing images using generative adversarial networks (GANs). The discriminator is transformed into a multi-class classifier, while the generator is used solely to assist in training the discriminator. Paper by Jain and Dhar (2017), the Inception V3 model is utilized for content-based image suggestions, and transfer learning is applied to develop neural networks.

Papers by Howard et al. (2017); Singh and Misra (2015, 2017); Zhang et al. (2017) introduce a creative approach for automatically categorizing diseases on plant leaves through the use of image segmentation methodology. Here, image segmentation serves as the key step for identifying the disease, which is accomplished using a genetic algorithm. The study also offers a comprehensive overview of the various techniques available for classifying plant leaf diseases, facilitating their identification (Filev et al., 2014; Zhang and Zhang, 2010).

6.3 PROPOSED METHODOLOGY

To execute our work in this paper, we have considered the following steps: dataset collection and preparation, training using a convolutional neural network (CNN) model for recognizing the crop type, training the CNN for disease detection, and validating the model based on various findings.

6.3.1 DATASET

Plant Disease Expert, an open-source dataset from Kaggle (n.d.) containing 60,129 images of crop leaves divided into 41 categories, is used. The dataset includes 29 different disease categories and 17 different plant species. Each class contains two fields specifying the names of the plant and the disease. Before any additional processing, all images are segmented, resized to 224 × 224 pixels, and converted into grayscale images.

6.3.2 IMAGE PREPROCESSING

The images in the dataset feature uneven illumination, which affects the model's accuracy. Image segmentation and noise removal during preprocessing are crucial for enhancing the quality of the CNN model. Segmentation is performed to extract only the relevant portion of the image. After segmentation, all the leaf images with correct brightness levels are obtained. Additionally, segmented images are converted to grayscale and sent for further processing to manage uneven illumination conditions (Kaggle, n.d.). Figure 6.1 displays the images obtained after preprocessing the strawberry leaves from the dataset.

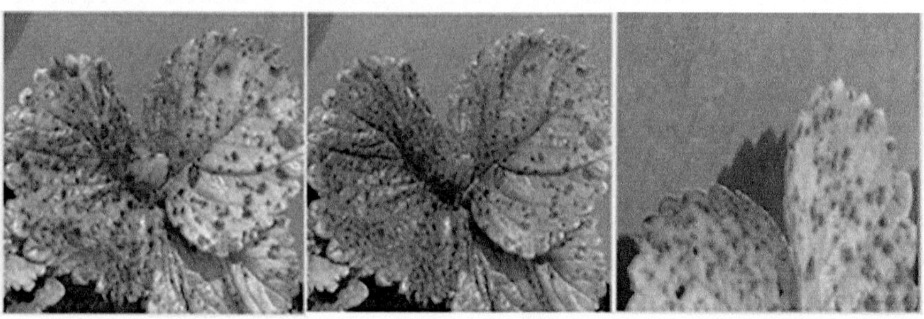

FIGURE 6.1 Plant leaf preprocessing – (i) original; (ii) grayscale; and (iii) segmented (brightness).

6.3.3 DETECTING AND CLASSIFYING PLANT DISEASE

Disease classification is carried out in the following steps: first, identifying the plant type, and second, identifying the disease type. These tasks are performed using deep CNNs. The ImageNet dataset is used to train the model through a deep learning (DL) process known as transfer learning process (TLP). In transfer learning, an existing model trained on one task is applied to another related task.

To tackle challenges efficiently and sustainably, we propose a method that develops neural networks for analogous tasks using a pretrained model. These pretrained networks are built through training on large-scale datasets (Big Data) containing a wide variety of image types. Such models are created by research organizations and require extensive time and computational resources using cutting-edge hardware. Once trained, these models are made available for reuse and can be incorporated into new models for similar problems.

The pretrained models are enhanced using the characteristics of the new dataset as well as the original training dataset. In such cases, the top layers of the network are retrained, and the tuned model is immediately deployed to address the task. If the existing dataset is sufficiently large, the model can be fully retrained with fresh data. In this scenario, the neural network initializes its weights using the pretrained model.

Utilizing transfer learning among a variety of established models such as ResNet50, MobileNet, VGG16, VGG19, Xception, and so on, we employ InceptionV3 and MobileNet for our implementation (ResearchGate, n.d.). As seen in Figure 6.2, InceptionV3 comprises 23,851,784 parameters and has a depth of 159 layers, whereas MobileNet contains 4,253,864 parameters and operates at a depth of 88 layers.

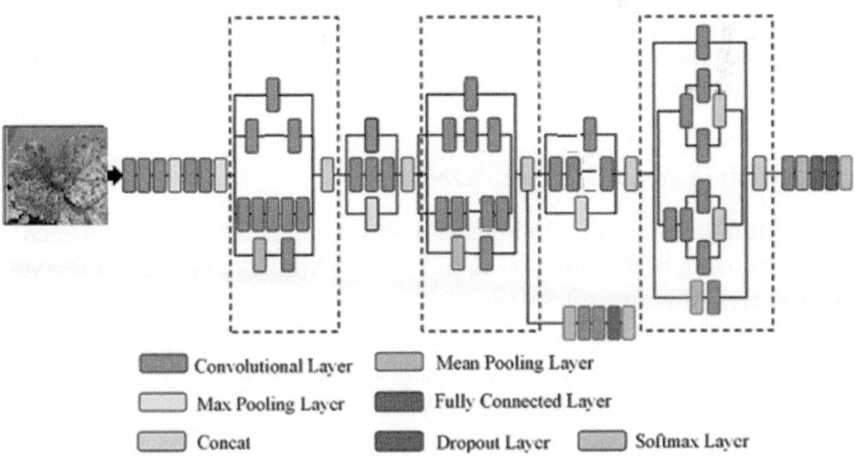

FIGURE 6.2 InceptionV3 architecture.

We have used an existing dataset of 6000 photos from Kaggle featuring 12 different types of plants, which has been used to fine-tune the newly developed models. Furthermore, using the weights from the previously trained model, the new models are initialized. The dataset is then preprocessed and divided in an 80:20 ratio for training and testing data, respectively. The technique of encoding labels is used to transform the output into categorical format.

An image data generator is utilized to diversify the input photos, thereby enriching the dataset. A dense layer equipped with the softmax activation function is introduced to the model for producing the final output. The Adam optimizer is employed, and categorical cross-entropy is used as the loss function. Re-training is carried out for 12 epochs using a batch size of 10 to fine-tune the previously trained model and produce the final model used in this paper. A dropout rate of 1e-3 is introduced to avoid the problem of overfitting. This refined model is tested using the testing dataset to verify its validation performance. The performance of both models is assessed based on training accuracy and loss, as well as validation accuracy and loss per epoch. These metrics provide insights into the models' effectiveness and generalization capabilities.

To detect disease specific to a particular plant, a deep CNN is employed for plant recognition techniques deployed by us. For disease identification, various models are created, each corresponding to a specific plant variety. The model first determines the type of plant and then selects the appropriate model for disease detection, which is used to analyze the image and identify the type of disease present. The disease detection models are evaluated using testing datasets to gauge their performance. This evaluation includes tracking metrics such as training accuracy, validation accuracy, training loss, and validation loss per epoch.

6.4 RESULTS AND DISCUSSION

This section discusses the findings and observations from the experiments conducted using both models (Mallick, 2015; Mallick et al., 2015; Nayak et al., 2021; Rautaray and Swain, 2021; Rout, Misra, and Samanta, 2014; Samanta et al., 2022). The segmented images model outperformed the color and grayscale image models once the trained model was put to the test. Both MobileNet and InceptionV3 models perform well in experiments for plant type detection, with accuracy rates of 99.7% and 99.77%, respectively (Figures 6.3–6.6).

The initial phases of accuracy growth show a significant increase, which later converges. Rapid initial learning is indicated by an exponential reduction in the loss function. It has been found that the InceptionV3 model outperforms MobileNet for plant identification.

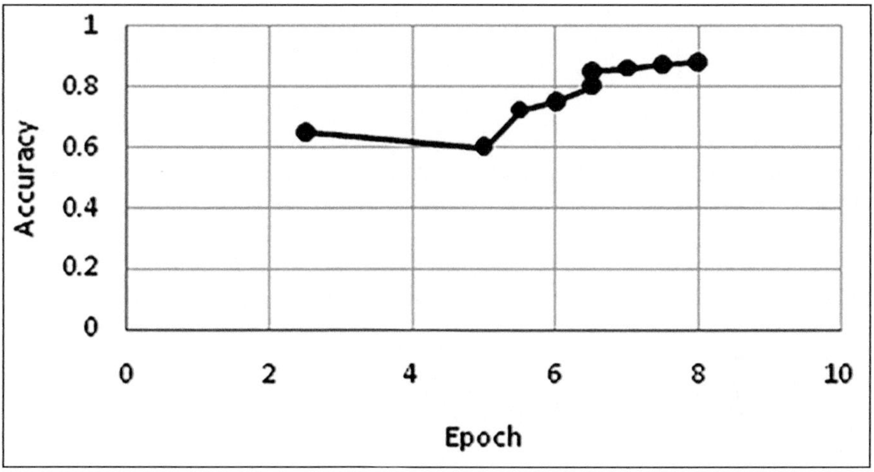

FIGURE 6.3 MobileNet: Plant identification – accuracy versus epoch.

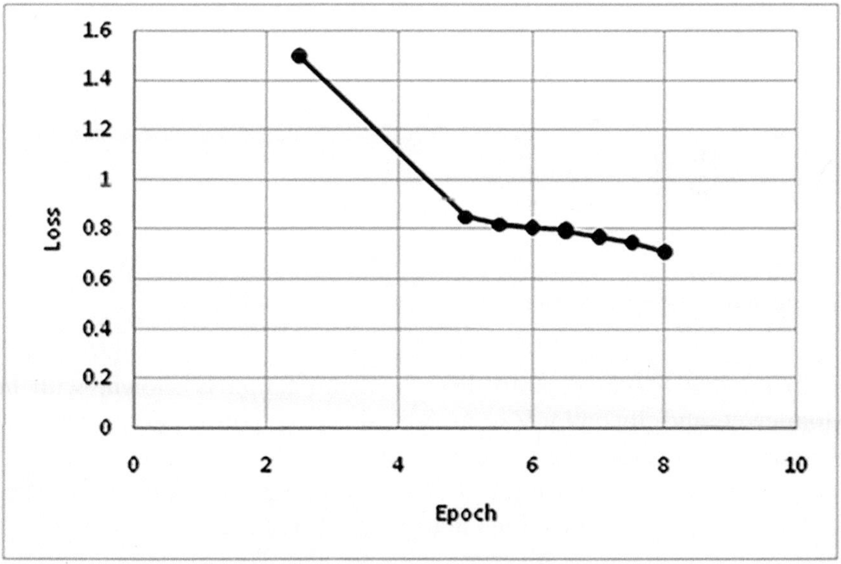

FIGURE 6.4 MobileNet: Plant identification – loss versus epoch.

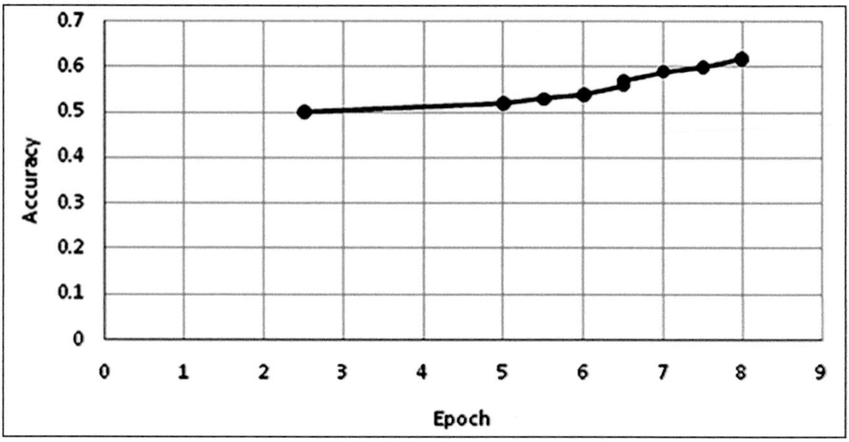

FIGURE 6.5 InceptionV3: Plant identification – accuracy versus epoch.

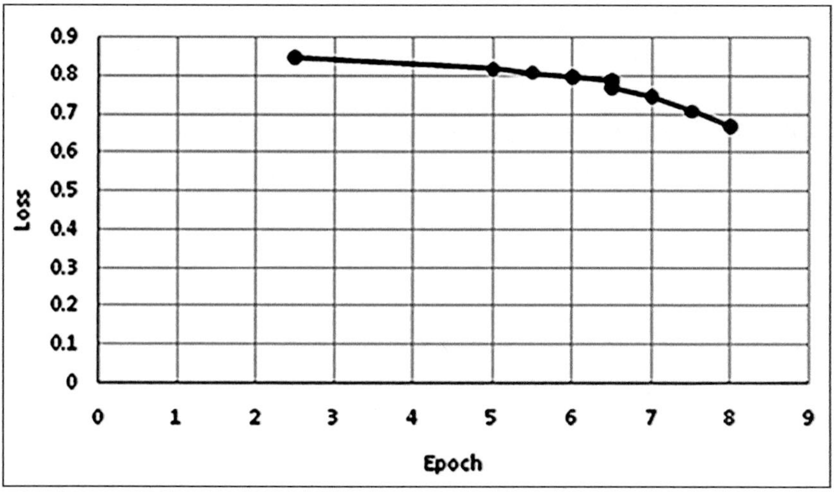

FIGURE 6.6 InceptionV3: Plant identification – loss versus epoch.

In Figures 6.7–6.10, crop disease identification is shown, with the MobileNet and InceptionV3 models showing consistent growth and accuracy rates, respectively. Between the seventh and eighth epochs, the InceptionV3 model's accuracy shows a very minor drop. The optimized performance of InceptionV3 in identifying disease is evidenced by the loss function value after the 12th epoch is completed. This suggests its efficacy in accurately classifying diseases compared to other models.

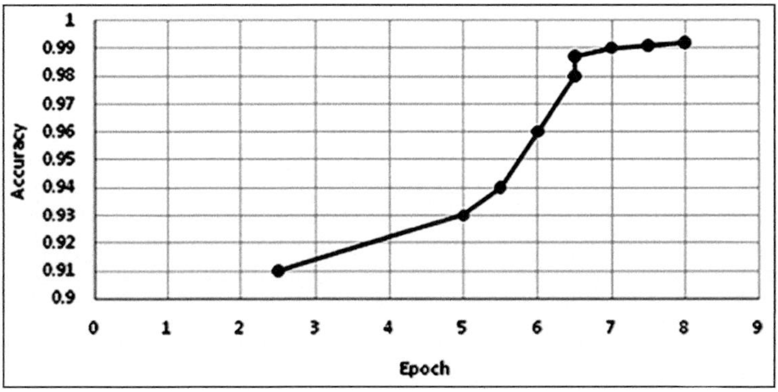

FIGURE 6.7 MobileNet: Disease identification – accuracy versus epoch.

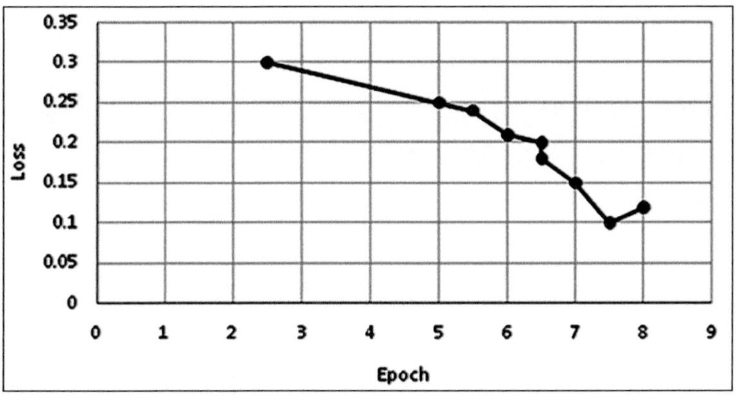

FIGURE 6.8 MobileNet: Disease identification – loss versus epoch.

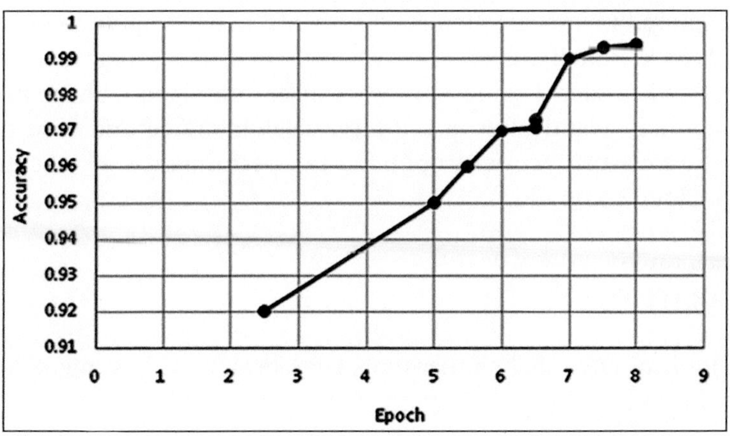

FIGURE 6.9 InceptionV3: Disease identification – accuracy versus epoch.

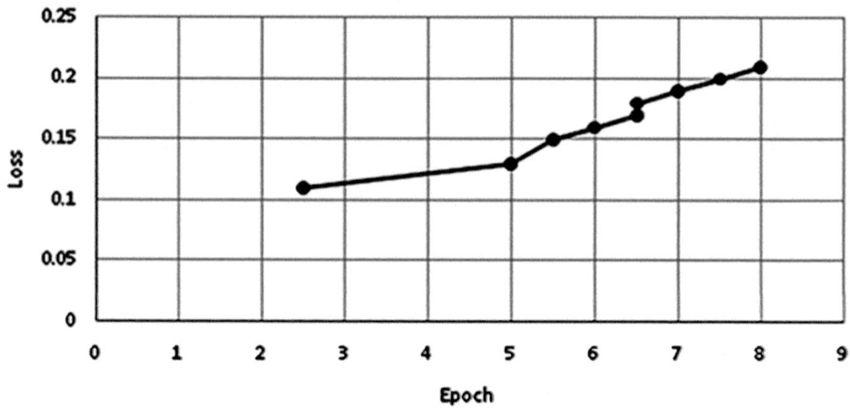

FIGURE 6.10 InceptionV3: Disease identification – loss versus epoch.

6.5 CONCLUSION

This research has explored applications of deep CNN in Python with the explicit aim of categorizing crop variety and identifying disease in plant pictures. The proposed methodology has been simulated using the Anaconda environment from samples of three distinct types of plant diseases and five different plant categories.

Based on experimental results, it is evident that the InceptionV3 model outperforms the existing MobileNet model when compared for parameters like accuracy and validation loss. Moving forward, this work will extend to categorizing images with diverse orientations and those captured in uncontrolled environments, expanding the scope of the research. Additionally, the variety of crop classes and their diseases can be expanded further. This technology can be combined with smartphone programs that would offer an intuitive user interface and ease of use. In our future work, we will optimize the performance of InceptionV3 through cloud computing.

DECLARATION

No AI tools or any other similar tools have been used in content creation of this chapter.

KEYWORDS

- **convolutional neural network**
- **crop destruction**
- **deep learning**
- **food security**
- **grey level co-occurrence matrix**
- **InceptionV3**
- **MobileNet**
- **plant disease**
- **support vector machine**
- **transfer learning**

REFERENCES

Arivazhagan, S., Shebiah, R. N., Ananthi, S., & Varthini, S. V. (2013). Detection of unhealthy region of plant leaves and classification of plant leaf diseases using texture features. *Agricultural Engineering International: CIGR Journal, 15*(1), 211. http://www.cigrjournal.org (accessed on 11th July 2025).

Filev, D., Jabłkowski, J., Kacprzyk, J., Krawczak, M., Popchev, I., Rutkowski, L., Sgurev, V., Sotirova, E., Szynkarczyk, P., & Zadrożny, S. (2014). *Advances in Intelligent Systems and Computing*. In *Proceedings of the 7th IEEE International Conference Intelligent Systems IS'2014*, Warsaw, Poland (Vol. 2): Tools, Architectures, Systems, Applications.

Howard, A. G., Zhu, M., Chen, B., Kalenichenko, D., Wang, W., Weyand, T., Andreetto, M., & Adam, H. (2017). MobileNets: Efficient convolutional neural networks for mobile vision applications. *arXiv preprint*, arXiv:1704.04861v1. https://arxiv.org/abs/1704.04861 (accessed on 11th July 2025).

Jain, S., & Dhar, J. (2017). Image-based search engine using deep learning. In *2017 Tenth International Conference on Contemporary Computing (IC3)*.

Kadir, A. (2014). A model of plant identification system using GLCM, lacunarity, and Shen features. *Research Journal of Pharmaceutical, Biological, and Chemical Sciences, 5*(2).

Kaggle. (N/A). Plant disease expert dataset. https://www.kaggle.com/datasets/sadmansakibmahi/plant-disease-expert (accessed on 11th July 2025).

Mallick, P. K. (2016). Research advances in the integration of big data and smart computing. IGI Global. https://doi.org/10.4018/978-1-4666-8737-0.

Mallick, P. K., Satapathy, B. S., Mohanty, M. N., & Kumar, S. S. (2015). Intelligent technique for CT brain image segmentation. In *2015 2nd International Conference on Electronics and Communication Systems (ICECS)* (pp. 1269–1277). IEEE.

Naik, M. R., & Sivappagari, C. (2016). Plant leaf and disease detection by using HSV features and SVM. *International Journal of Engineering Science and Computing (IJESC), 6*(12).

Nayak, S. R., Sivakumar, S., Bhoi, A. K., Chae, G. S., & Mallick, P. K. (2021). Mixed-mode database miner classifier: Parallel computation of graphical processing unit mining. *The International Journal of Electrical Engineering & Education*. https://doi.org/10.1177/0020720920988494 (accessed on 11th July 2025).

Olmschenk, G., Tang, H., & Zhu, Z. (2018). Crowd counting with minimal data using generative adversarial networks for multiple target regression. In *2018 IEEE Winter Conference on Applications of Computer Vision (WACV)*.

Rautaray, B., & Swain, D. K. (2021). Intellectual structure and publication pattern in *Journal of Money Laundering Control* from 2014 to 2020: A bibliometric study.

ResearchGate. (N/A). The architecture of Inception V3 model. https://www.researchgate.net/figure/The-architecture-of-Inception-V3-model_fig5_349717475 (accessed on 11th July 2025).

Rout, S. S., Misra, B. B., & Samanta, S. (2014). Load allocation in academic environment: A multi-objective PSO approach. *GSTF Journal on Computing (JoC), 3*(4), 1–12.

Samanta, S. R., Mallick, P. K., Pattnaik, P. K., Mohanty, J. R., & Polkowski, Z. (2022). *Cognitive Computing for Risk Management*. Springer.

Singh, V., & Misra, A. K. (2015). Detection of unhealthy region of plant leaves using image processing and genetic algorithm. In *2015 International Conference on Advances in Computer Engineering and Applications (ICACEA)*. IEEE.

Singh, V., & Misra, A. K. (2017). Detection of plant leaf diseases using image segmentation and soft computing techniques. *Information Processing in Agriculture, 4*. Elsevier.

SVM classifier based grape leaf disease detection. (2016). In *2016 Conference on Advances in Signal Processing (CASP). Agricultural Engineering International: CIGR Journal, 15*(1), 211. IEEE. http://www.cigrjournal.org (accessed on 11th July 2025).

Zhang, C., Zhang, S., Yang, J., Shi, Y., & Chen, J. (2017). Apple leaf disease identification using genetic algorithm and correlation-based feature selection method. *International Journal of Agricultural and Biological Engineering, 10*(2). https://www.ijabe.org (accessed on 11th July 2025).

Zhang, J., & Zhang, W. (2010). Support vector machine for recognition of cucumber leaf diseases. IEEE.

CHAPTER 7

Performance Analysis of Various Machine Learning Techniques for Fruit Classification and Counting from Images

BISHAL PRADHAN,* RAJKIRAN UPADHYAY, and SAMARJEET BORAH

Department of Computer Applications, Sikkim Manipal Institute of Technology, Sikkim Manipal University, Sikkim, India

*Corresponding author

ABSTRACT

The concept of precision agriculture focuses on sustainability and optimal resource utilization. Computer vision is a field that enables a computer to understand and interpret the world through images and video. It has been extensively studied in various aspects of precision farming, including autonomous harvesting robots in plant phenotyping, harvesting, detection of plant pests and diseases, pricing, and more. Computer vision is being used to detect fruits and vegetables to determine their three-dimensional position. It can also be used to detect plants, pests, diseases, and so forth. With the development of computer vision, such technologies have been widely adopted in the field of agricultural automation. In the future, technologies like computer vision and intelligent systems such as deep learning (DL) can be applied to each facet of agriculture.

Intelligent Computing and Analytics in Image Processing. Samarjeet Borah, Jambi Ratna Raja Kumar, Laxmi Bewoor, Rahul Raghvendra Joshi, and Mahendra Prabhakar Deore (Eds.)

DOI: 10.1201/9781779640772-7

7.1 INTRODUCTION

Precision agriculture refers to the concept of agricultural management based on the observation, measurement, and response to variability in plants. Latest technologies complement farmers' observations and add an additional dimension to the evaluation of agricultural system performance. It is found that progress in agriculture is largely based on observations and variability in crops, whether through images or physical inspection in the field.

Precision agriculture is a newly adopted concept aimed at increasing production, reducing labor time, and ensuring effective agricultural management. The impact of artificial intelligence (AI) on all sectors is growing steadily. As computer vision and deep learning (DL) continue to advance, they can handle increasingly complex tasks with greater ease. Agriculture is a sector that can benefit significantly from technology. With several countries not meeting supply requirements, it is important to adopt technologies that improve production and overall efficiency. Computer vision is proving to be a significant advancement in agriculture. Using computer vision and DL concepts reduces costs for farmers on a large scale. For example, DL can be applied to classify fruits, and computer vision can count the number of fruits, enabling timely intervention. The use of computer vision and DL algorithms in agriculture is expected to rise, due to the complexity of the ground and climate in which they operate. In the future, computer vision and intelligent technologies like DL will be applied to every aspect of agriculture.

7.1.1 PROBLEMS IN FRUITS CLASSIFICATION

There are several issues in the classification of fruits through image processing. Some of these are briefly stated below:

1. **Intra-Class Variation:** This refers to the variation between images of the same class.
2. **Scale Variation:** Scale variation occurs when an image of the same object appears in multiple sizes. It is a common problem in image classification.
3. **Viewpoint Variation:** This occurs when an object is rotated in various dimensions. Regardless of the angle at which the image of a fruit is captured, it remains the same fruit (e.g., a mango).

4. **Illumination:** Variation in intensity levels.
5. **Overfitting:** This occurs when a function closely fits the training data, even learning noise as concepts. These learned concepts do not generalize to unseen data, resulting in decreased model performance on new inputs.
6. **Training Time:** DL requires significant training time. Models with many hidden layers and large datasets demand substantial computational power, making them slow and resource-intensive. Training time increases with network complexity.

7.1.2 PROBLEMS IN FRUITS COUNTING

Similarly, fruit counting through image processing also faces several challenges, as stated below:

1. **Occlusion:** This occurs when the object to be identified is not fully visible or is hidden behind another object.
2. **Illumination and Background Clutter:** Variable intensity levels and noisy images lead to incorrect foreground and background segmentation. In some cases, a specific object appears visually similar to the background, making it difficult for the fruit counting system to detect it.
3. **Overlapping Objects:** When objects overlap, it becomes extremely difficult to distinguish between them. Therefore, fruit counting systems must be able to handle overlapping objects to reliably count the number of fruits.

7.1.3 AIM AND OBJECTIVE OF THE WORK

The aim of this work is the classification of fruits using DL techniques, where the convolutional neural network (CNN) and various other algorithms are tested to correctly classify fruits given their images and to count the number of fruits using computer vision techniques. The task is to train the deep neural network in a supervised manner, where fruit images will be the input of the network and fruit labels will be the output. After successful training, the various models used will be able to accurately predict the fruit class.

The objectives are to compare, analyze, and determine various techniques and models in terms of accuracy, speed, and reliability for fruit classification and counting. This includes evaluating the quality of the classifier by requesting it to predict labels for an unseen set of images and comparing the predicted labels with the true labels. The goal is to improve the system's ability to classify images with greater accuracy. In addition, the objective includes processing images to overcome challenges using appropriate techniques. These challenges may be the primary barriers to better results in fruit counting, such as overlapping objects, occlusion, lighting variations, background changes, and so forth.

7.2 RELATED WORKS

Several influential research works have been reported in the literature on fruit estimation and counting, which is a part of precision farming. Some of the works are discussed in this section.

A comparative review is presented by Sayal, Garg, & Sharma (2013), covering various techniques for L×a×b color space transformation of RGB images, which provides better segmentation results using K-means clustering or color-based K-nearest neighbor (KNN) classifiers. A CNN-based classifier model with a hidden layer of 256 nodes has been proposed by Asriny, Rani, and Hidayatullah (2020) to classify orange images. As stated, the model is validated using the K-Fold Cross Validation method, and two activation functions—ReLU and Tanh—have been compared with respect to the softmax classifier. The accuracy of the ReLU activation function is found to be better than that of the Tanh activation function.

A detailed comparison of various image processing and machine learning (ML) technologies for disease detection and classification techniques related to agricultural produce is discussed by Ganatra and Pate (2018). Additionally, various color representation schemes, feature extraction methods, and segmentation techniques, along with their pros and cons, have been reviewed. A method for green fruit detection using feature extraction in digital images of orange trees has been proposed by Maldonado and Barbosa (2016), showing a detection rate with false positives (FPs) of 3% in images acquired under favorable conditions.

An extensive comparative study has been presented by Mavridou et al. (2019) regarding the use of computer vision techniques in various aspects of precision farming, mainly in three functional categories: (i) plant and fruit detection approaches; (ii) grading, maturity, counting, and yield prediction; and (iii) plant and fruit health monitoring. A simple fruit classification and recognition system has been proposed by Seng & Mirisaee (2009), which combines three features—color, shape, and size—using the KNNs classification technique. It is found to be 90% accurate in classifying and identifying fruits.

A method for the automatic quantification of apple videos is presented by Bhuyan and Borah (2014). This method relies on the implementation of the YOLOv4-tiny detection network and a single-object tracking algorithm. It aims to accurately assess fruit quantity and estimate yield in apple orchards by combining detection, tracking, and counting.

More image processing-related works are available in the literature (Bhuyan and Borah, 2014; Borah et al., 2018; Kalita et al., 2018). Similar works can also be found in Bhuyan and Borah (2014); Kumar et al. (2014).

7.3 RESEARCH METHODOLOGY

In this work, the task is performed in two phases: feature extraction and classification. In the first phase, CNN and VGG16 algorithms are used to perform feature extraction, followed by four selected classification algorithms—XGBoost, random forest, artificial neural network (ANN), and support vector machine (SVM).

The convolutional layer is responsible for feature extraction. Max pooling (2×2) is used to reduce the spatial size of the feature map after convolution. It also helps in significantly reducing the computational cost. By reducing the size of the feature map, we effectively reduce the computational load. The convolution and pooling operations may be repeated several times, depending on the problem statement. The flatten layer is used to convert a multi-dimensional array into a one-dimensional (1D) array so that inputs can be fed to neurons individually for further operations.

Convolution and pooling operations are repeated multiple times to extract various types of features. After these operations, the output from convolution is converted from a two-dimensional (2D) array to a one-dimensional (1D) flattened array. Neurons in the hidden and output layers are connected to each other to create a complex architecture. The first convolutional layer acts as an edge detector. After the second convolution, the resultant bright areas represent the activated regions. Various types of filters are responsible for extracting different features during convolution, with each filter capturing unique information. In the third convolution, the feature maps become sparser, meaning the filters detect fewer features, and the vital information cannot be seen in one place. Deeper feature maps contain less detailed information about the image but more abstract information relevant to the image's class.

The experimental mechanisms considered in this work are as follows:

- CNN as feature extractor and random forest as classifier.
- VGG-16 as feature extractor and XGBoost as classifier. VGG-16 is a modified version of CNN. The number 16 refers to the 16 layers in the network.
- CNN as feature extractor and SVM as the final layer. SVM is a supervised algorithm typically used for both regression and classification problems.

The data points, initially in two-dimensional space, are projected into another dimension so that the points belonging to different classes can be separated clearly. This is achieved by drawing a hyperplane that separates the data points of different classes effectively and with ease.

7.4 IMPLEMENTATION

The experiment was conducted on an Intel i5-based machine with 8 GB of RAM. Python was used as the programming language on a Windows platform. The steps for fruit counting are explained below:

- **Step 1:** Collect raw images from sources.
- **Step 2:** Convert RGB to HSV.
- **Step 3:** Convert grayscale image into black and white (binarize) using thresholding.
- **Step 4:** Remove noise from the image.
- **Step 5:** Adjust the size of the object using morphological operations such as erosion, dilation, opening, and closing.
- **Step 6:** Provide a metric or measure of the separation of points in the image (distance transform).

- **Step 7:** Define a template to extract a small image of a targeted region similar to a template image.
- **Step 8:** Perform template matching on the distance-transformed image.
- **Step 9:** Apply thresholding to the distance-transformed image.
- **Step 10:** Draw contours to identify the structure of the object.

The results obtained from the experiments are discussed in the next section.

7.5 RESULTS AND DISCUSSION

7.5.1 *FRUITS COUNTING*

The experiment is conducted based on publicly available fruit pictures. In this section, source images along with the results are presented. The resultant images are directly copied from the Jupyter Notebook (Figures 7.1–7.10). In Figure 7.1, we can see two oranges in the source image, which are distance transformed to highlight the central regions of the fruits for further analysis or counting.

FIGURE 7.1 Simple image for fruit counting.

```
Number of Oranges :  2
<matplotlib.image.AxesImage at 0x194fc78aa90>
```

FIGURE 7.2 Output for fruit counting.

FIGURE 7.3 Input image for fruit counting from dense image.

```
Number of oranges :  77

<matplotlib.image.AxesImage at 0x17b42e4a6d0>
```

FIGURE 7.4 Output for fruit counting from dense image.

FIGURE 7.5 Output for fruit counting from dense image.

```
Number of apples :  41

<matplotlib.image.AxesImage at 0x16301880a30>
```

FIGURE 7.6 Output for fruit counting from dense image.

FIGURE 7.7 Input image for fruit counting with similar background.

```
Number of Mangoes :  163

<matplotlib.image.AxesImage at 0x1e04dbbf
```

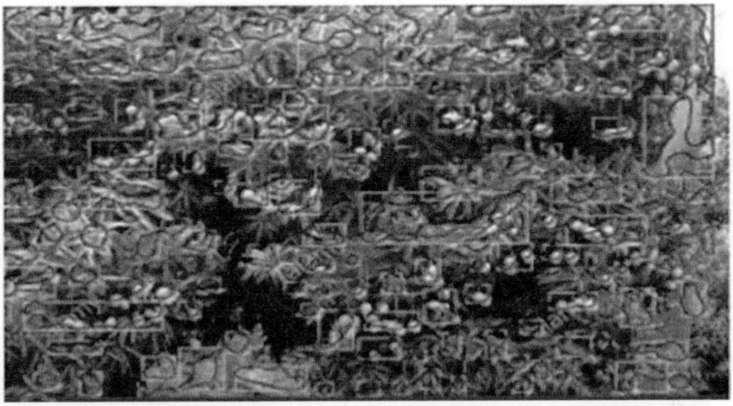

FIGURE 7.8 Output for fruit counting from image with similar background.

FIGURE 7.9 Input for unseen test data.

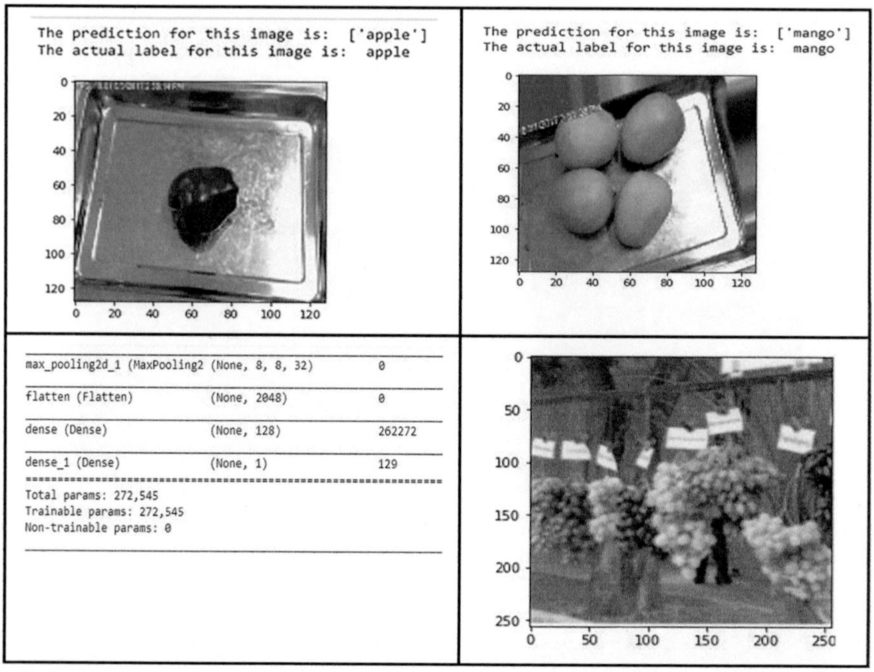

FIGURE 7.10 Output for unseen test data.

7.5.2 FRUITS CLASSIFICATION

The experiment has been conducted for both binary and multiclass classification. Both outcomes are presented in subsections.

7.5.2.1 Binary classification

In this experiment, the ratio for training and validation is set to 70:30. There are four different executions with different combinations of algorithms, such as:

- VGG-16 as feature extractor and XGBoost as the classifier;
- CNN as feature extractor and Random Forest as the classifier;
- CNN as feature extractor and ANN as the classifier;
- CNN as feature extractor and SVM as the classifier.

The outcomes of the experiment are represented in Table 7.1. It is found that the accuracy obtained by using the Random Forest classifier is highest, but the SVM trains fastest. A graphical representation of the same is given in Figure 7.11.

TABLE 7.1 Binary Classification Analysis

Algorithm	VGG-16 and XGBoost	CNN and Random Forest	CNN and ANN	CNN and SVM
Accuracy	90%	94%	93%	88%
Training Time	100 minutes	80 minutes	70 minutes	33 minutes
Epochs	–	–	85	53

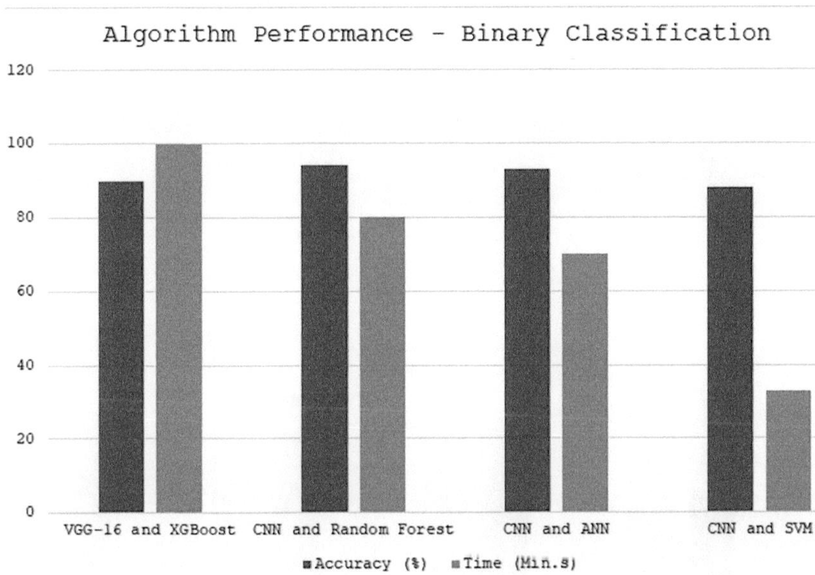

FIGURE 7.11 Performance of the selected algorithms on binary classification.

7.5.2.2 Multi-class classification

In the multi-class classification experimental setup, the proportion allocated for training and validation is fixed at 70% and 30%, respectively. Four distinct runs were conducted utilizing various algorithmic configurations (as in the previous execution):
VGG-16 as feature extractor and XGBoost as the classifier;

- CNN as feature extractor and Random Forest as the classifier;
- CNN as feature extractor and ANN as the classifier;

- CNN as feature extractor and SVM as the classifier.

The results of the experiment are shown in Table 7.2. The XGBoost classifier produces the maximum accuracy, although SVM trains quickly. Figure 7.12 shows a graphical illustration of the same.

TABLE 7.2 Multi-Class Classification Analysis

Algorithms	VGG-16 and XGBoost	CNN and Random Forest	CNN and ANN	CNN and SVM
Accuracy	98%	85%	97%	89%
Training Time	125 minutes	114 minutes	180 minutes	55 minutes
Epochs	–	–	96	30

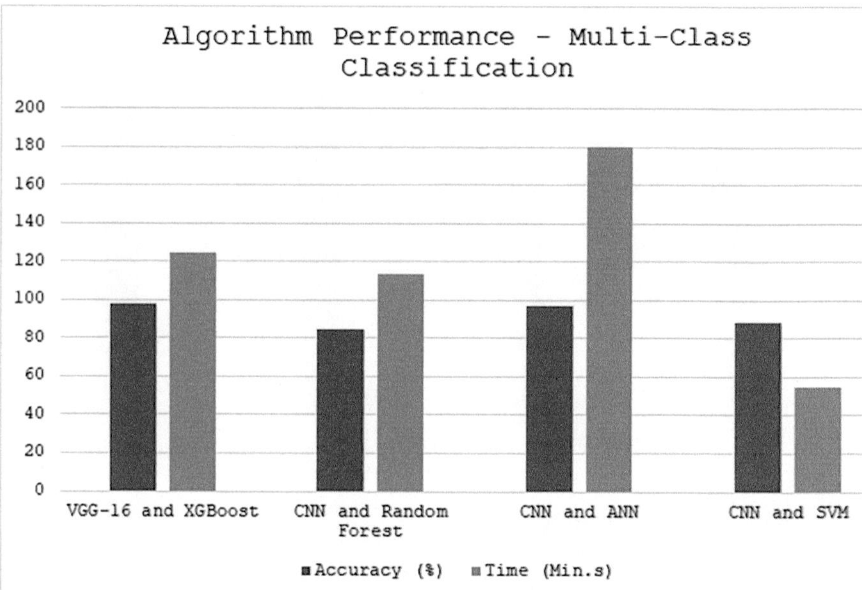

FIGURE 7.12 Performance of the selected algorithms on multi-class classification.

7.6 CONCLUSION

In this research work, VGG-16, XGBoost, CNN, Random Forest, and ANN are used to conduct a comparative analysis on performance. Various combinations such as VGG-16 & XGBoost, CNN & Random Forest, CNN & ANN, and CNN & SVM are used as feature extractor and classifier, subsequently. The models are fast, precise, and affordable to adopt easily, offering high efficiency, and the addition of many useful features in the system will make things easier for the users of this system. It is found from the research work that fruits are being classified successfully, as well as the number of fruits on a tree being counted. It is found that computer vision techniques have some limitations in separating the objects when background and foreground intensity levels are almost the same.

However, there are multiple scopes for improvement, such as the use of DL techniques for fruit counting.

DECLARATION

No AI tools or any other similar tools have been used in content creation of this chapter.

KEYWORDS

- **algorithmic configurations**
- **big data**
- **binary classification**
- **computer vision techniques**
- **deep learning techniques**
- **machine learning**
- **neural networks**
- **precision agriculture**

REFERENCES

Asriny, D. M., Rani, S., & Hidayatullah, A. F. (2020). Orange fruit images classification using convolutional neural networks. *IOP Conference Series: Materials Science and Engineering, 803*, 012020. https://doi.org/10.1088/1757-899X/803/1/012020 (accessed on 11th July 2025).

Bhuyan, R., & Borah, S. (2014). Image segmentation: Computational approaches for medical images. *International Journal of Computer Applications (IJCA), Special Issue on Bioinformatics and Computational Biology,* (3), 13–17. https://www.ijcaonline.org/proceedings/icbcb/number3/17629-1503 (accessed on 11th July 2025).

Borah, S., Pradhan, R., Dey, N., & Gupta, P. (2018). Soft Computing Techniques and Applications. *Advances in Intelligent Systems and Computing*. Springer. ISBN: 978-981-13-1279-3.

Ganatra, N., & Pate, A. (2018). A survey on diseases detection and classification of agriculture products using image processing and machine learning. *International Journal of Computer Applications, 180*(13). https://doi.org/10.5120/ijca2018916265 (accessed on 11th July 2025).

Gao, F., Fang, W., Sun, X., Wu, Z., Zhao, G., Li, G., Li, R., Fu, L., & Zhang, Q. (2022). A novel apple fruit detection and counting methodology based on deep learning and trunk tracking in modern orchard. *Computers and Electronics in Agriculture, 197*, 107000.

Kalita, J., Balas, V. E., Borah, S., & Pradhan, R. (2018). Recent developments in machine learning and data analytics. *Advances in Intelligent Systems and Computing*. Springer. ISBN: 978-981-13-1279-3.

Kumar, I., Jha, R. S., Kumar, S., & Borah, S. (2014). Brain abnormality detection from MR images using matrix symmetry method. *International Journal of Computer Applications (IJCA) Special Issue on Bioinformatics and Computational Biology,* (3), 22–25. https://www.ijcaonline.org/proceedings/icbcb/number3/17631-1505 (accessed on 11th July 2025).

Maldonado, W. Jr., & Barbosa, J. C. (2016). Automatic green fruit counting in orange trees using digital images. *Computers and Electronics in Agriculture, 127*, 572–581.

Mavridou, E., Vrochidou, E., Papakostas, G. A., Pachidis, T., & Kaburlasos, V. G. (2019). Machine vision systems in precision agriculture for crop farming. *Journal of Imaging, 5*(12), 89. https://doi.org/10.3390/jimaging5120089 (accessed on 11th July 2025).

Sayal, A., Garg, D., & Sharma, S. (2013). A survey of computer vision methods for counting fruits and yield prediction. *International Journal of Computer Science Engineering, 2*(6).

Seng, W. C., & Mirisaee, S. H. (2009). A new method for fruits recognition system. In *2009 International Conference on Electrical Engineering and Informatics*, Bangi, Malaysia (pp. 130–134). IEEE. https://doi.org/10.1109/ICEEI.2009.5254804 (accessed on 11th July 2025).

CHAPTER 8

An Analysis of Image Edge Detection Techniques for Forensics Analysis

VITTHAL KAMBLE[1,*] and NILESH UKE[2]

[1]*Department of Computer Engineering, BRACT's Vishwakarma Institute of Information Technology, Pune, Maharashtra, India*

[2]*Research Supervisor, Department of Computer Engineering, Vishwakarma Institute of Information Technology, Affiliated to Savitribai Phule Pune University (SPPU),Pune, Maharashtra, India*

*Corresponding author

ABSTRACT

Image edge detection is an essential piece of information these days. Especially in medical image analysis, edge detection can provide realistic facts. The rigid areas of the photo with variance are identified by image contour detection techniques. When considerable and quick changes take place in the intensity of the image, it results in the formation of edges. These changes occur in the inner frame of the object in an image, where a sudden intensity change differentiates the object from its environment. Using edge detection techniques, one can find curves and lines within an image.

The applications of edge detection include image analysis such as pattern recognition, feature selection, and object recognition. In forensic analysis, edge detection has significant impact in face recognition, fingerprint examination, anatomical study, medical imaging, and so forth. Through this article, we have examined edge detection techniques and outlined their importance and various applications.

Intelligent Computing and Analytics in Image Processing. Samarjeet Borah, Jambi Ratna Raja Kumar, Laxmi Bewoor, Rahul Raghvendra Joshi, and Mahendra Prabhakar Deore (Eds.)

DOI: 10.1201/9781779640772-8

8.1 INTRODUCTION

When considerable and quick changes take place in the intensity of the image, it results in the formation of edges. There are namely three types of edges: horizontal, vertical, and diagonal (Tagunde, Uke, and Banchhor, 2012). Contour detection technique is one of the ubiquitous methods for segmenting an image into its areas of discontinuity. It helps in simplifying the structure of an image and studying its features. It indicates the texture change in the gray level, marking the end of one texture and the start of a new one (Jain, 2008). In layman's terms, utilizing edge detection techniques allows the identification and precise location of sharp discontinuities in a photograph (Meshram and Lande, 2018; Uke and Thool, 2013).

The steps for photograph border detection are:

i. **Consider a Color Image:** Use a color image as the input for processing

ii. **Image Smoothening:** Removing noise in the images without altering their quality.

iii. **Enhance Image Quality:** Eliminate noisy pixels while preserving edge details to improve the overall image quality.

iv. **Estimate the Location of Edges:** Retain edges of the image, which are found by sudden intensity changes.

v. **Edge Localizing:** Linking the edges to locate them properly,

vi. **Output:** Obtain the output image.

- **Approaches for Border Detection:** With growing requirements for object detection, digital photograph processing, border detection, object recognition, 3D modeling, and image tracking, edge detection is essential (Meshram and Lande, 2018; Uke and Thool, 2013). Many algorithms have been created for photograph border detection.

8.2 LITERATURE SURVEY

A few significant works from the literature have been selected for study, and the outcomes are displayed in Tables 8.1 and 8.2.

TABLE 8.1 Summary of Review Works

Sl. No.	Author	Year	Type of Algorithm	Advantages	Disadvantages	Techniques
1.	Adi Budi Prasetyo, Rashi Wahyudi, Imam Tahyudin, Selvia Ferdiana Kusuma, Luzi Dwi Oktaviana, Azhari Shouni Barkah, Budi Artono	2023	Edge detection (Prewitt, Laplace, Kirsch operators)	The Prewitt operator produces the best edge detection.	Not mentioned	The MSE, PSNR, and image difference values are used for the analysis process.
2.	Unknown	2021	Edge detection (Sobel, Prewitt, Robert, Laplacian)	The technique is proposed to obtain thin edges.	It becomes strenuous to extract edges when there is noise in an image.	Comparative analysis for four distinct methods for detecting borders.
3.	Edy Winarno, Wiwien Hadikurniawati, Setyawan Wibisono, Anindita Septiarini	2021	Edge detection and grey level co-occurrence matrix (GLCM)	The characteristics of human fingers are used for fingerprint identification.	Not mentioned	GLCM method is used for preprocessing of edge detection and feature extraction.
4.	Unknown	2015	Edge detection (Robert's, Prewitt, Sobel, Canny)	Edge detection is used to extract important features.	Distorted image edge detection is tough.	The output of images is shown using the software MATLAB.
5.	Deepali Srivastava, Rashi Kohli, Shubhi Gupta	2017	Prewitt, Sobel, Robert	Disparate edge detection techniques are used.	Distorted image edge detection is tough.	Disparate analysis.

TABLE 8.2 Summary of Sobel, Prewitt, Roberts, Canny, and Fuzzy Logic Techniques

Sl. No.	Algorithm/ Technique	Advantages	Disadvantages	Performance
1.	Sobel	• Simple and time-efficient. • Easy to implement, fast to compute, and robust to noise. • Detects edges and their orientations.	• Produces rough edges. • Sensitive to noise. • Sometimes inaccurate.	Good
2.	Prewitt	• Similar to Sobel.	• Does not emphasize pixels closer to the center of the mask.	Good
3.	Roberts	• Simplest edge detector. • Quickly computes 2D gradient.	• Sensitive to noise. • Often inaccurate.	Good
4.	Canny	• Produces smoother, thinner, and cleaner edges. • Accurate edge localization. • Minimal response to noise.	• Computationally complex. • Time-consuming. • May produce false zero crossings.	Better
5.	Fuzzy logic	• Offers adaptability. • Easier to code than standard logical programming. • Requires fewer instructions. • Saves memory. • Effective in detecting edges in smooth clinical images. • Performs well in smooth and noisy images.	• Sometimes inaccurate.	Better for medical/ forensic images

8.3 EDGE DETECTION TECHNIQUES

8.3.1 CANNY BORDER DETECTION

Canny border detection is a multistage process used to extract informative structural details from various visual elements. It reduces the amount of information to be processed. Its applications can be found in computer vision systems. It is simple to implement, has fast execution, a low error rate, and offers accurate edge localization. Its robustness to noise makes it a competitive option among other detection techniques.

Steps for canny border detection algorithm working:

- **Step 1:** Take any image and convert it to a grayscale image.
- **Step 2:** Smoothen the image using a Gaussian filter. Using this, the removable errors are eliminated.
- **Step 3:** Gaussian filter equation:
- **Step 4 (Meshram and Lande, 2018; Uke and Thool, 2013):**

$$G(X,Y) = G_{\delta}(X,Y) \times G(X,Y) \tag{1}$$

- **Step 5:** Where; $G_{\delta} = \frac{1}{\sqrt{2\pi\delta^2}} \exp\left(\frac{-X^2 + Y^2}{2\delta^2}\right)$
- **Step 6:** Following this, the gradient magnitude can be calculated through gradient operations:
- **Step 7 (Meshram and Lande, 2018; Uke and Thool, 2013):**

$$M(X,Y) = \sqrt{(g_X^2(X,Y) + g_y^2(X,Y)} \tag{2}$$

- **Step 8 (Meshram and Lande, 2018; Uke and Thool, 2013):**

$$\theta(X,Y) = \tan^{-1}\left[\frac{g_y(x,y)}{g_x(x,y)}\right] \tag{3}$$

- **Step 9:** Use of a differentiated Gaussian filter, based on the first derivative of pixel intensities and the normal Gaussian filter.
- **Step 10:** It produces an output that already highlights the edges in the image.
- **Step 11:** Output (Figures 8.1 and 8.2):

The high values indicate that the first derivative is high at that location. Edge normal is checked for high values. Later, only high values are retained, and others are set to 0, forming an edge.

8.3.1.1 Hysteresis thresholding

Prevents noisy edges from causing fragmentation. Two threshold values are considered: *Thresh1* and *Thresh2*, where *Thresh1* is greater. Non-maximum suppression (NMS) tracing will start if gradient values are higher than *Thresh1* and will stop if they fall below *Thresh2*.

FIGURE 8.1 Deer's original photograph processed using Canny border detection algorithm.

8.3.2 SOBEL EDGE IDENTIFICATION

The gradient operator is applied in this method. It operates on a 2D positional Sobel measurement over an image, thereby focusing on areas of fine detail with high spatial frequencies corresponding to corners of the image. It typically searches the calculated absolute Sobel intensity for every detail in a gradient photo image.

The new image formation is achieved by processing the original image in the *X* and *Y* directions, which is effectively the sum of the *X* and *Y* edge components of the image. The procedure is carried out by calculating the gradient of the image power for each bit in the picture (Fisher, 2004).

➢ **Working:**

1. A pair of 3×3 convolution kernels is used, that is, two 3×3 matrices. The first kernel is simply the 90° rotated version of the second kernel.

   The two gradient components are created as follows:

- ***X*-direction kernel (G_x):**

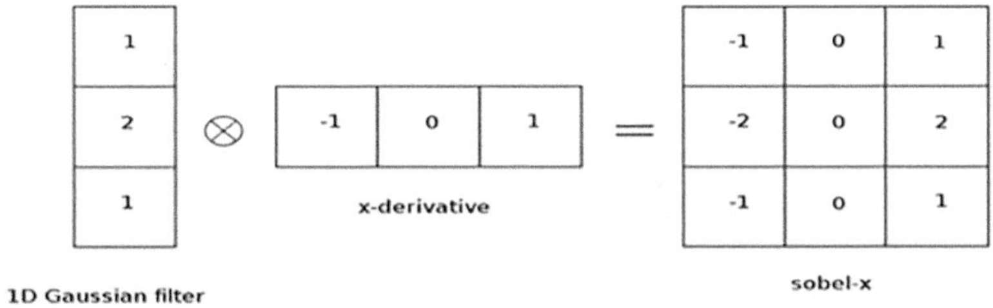

- ***Y*-direction kernel (G_y):**

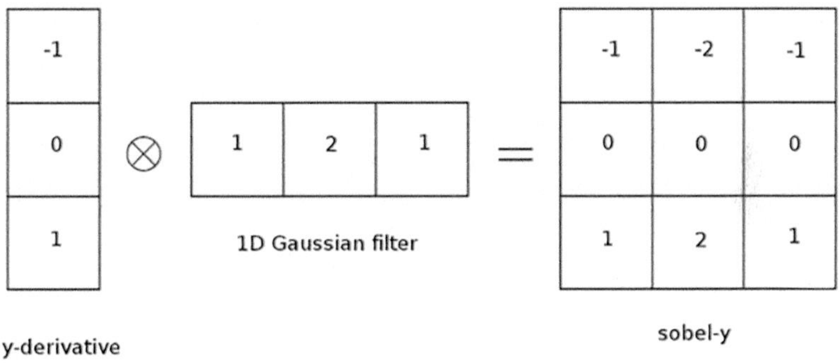

-1	0	+1
-2	0	+2
-1	0	+1

Gx

+1	+2	+1
0	0	0
-1	-2	-1

Gy

2. These two kernel masks are applied separately to the image to get individual gradient components for the Sobel operator in each direction (G_x and G_y).
3. These components are combined to compute the Sobel intensity and direction:

$$|G| = \sqrt{G_x^2 + G_y^2} \quad (4)$$

Alternatively, a simpler approximation can be used:

$$|G| = |G_x| + |G_y| \quad (5)$$

The angle representing the direction of the edge that causes spatial variation is calculated by:

$$\Theta = \arctan\left(G_y / G_x\right) \tag{6}$$

8.3.3 FUZZY LOGIC ALGORITHM

There have been many new applications based on the use of fuzzy logic, from brake systems in trains to washing machines. It also plays a significant role in image contour detection algorithms. The formula is established on the evaluation of four pixels within a 2×2 window of an image using a set of fuzzy conditions that help highlight all edges associated with that image (Chinchkhede and Uke, 2012). In the presence of an edge, the fuzzy logic conditions assist in evaluating the proportional values of the bits to be presented, thereby extracting all the edges.

P_1	P_2
P_3	P_4

2×2 mask used for scanning.

When a portion of the mask is slid over an input image, performance is achieved through a fuzzy inference framework that depends on fuzzy logic protocols and the pixel values within the mask.

- Inputs for the pixels are fuzzified using conventional functions—black and white.
- Fuzzy logic rules are applied to the input.
- A comprehensive result (FS) for all rules is obtained using the MAX operator (Chinchkhede and Uke, 2012).
- The centroid method is used for de-fuzzification.
- The bit of pixel P_4 becomes the result in the output image, showing the edges in Black and White regions
- Noise removal is then applied for refinement.

8.4 RESULTS AND DISCUSSIONS

The research explored three prominent edge detection techniques—Canny edge detection, Sobel edge identification, and a fuzzy logic algorithm—in the context of forensic and medical image analysis. Canny edge detection demonstrated accuracy with low error rates, Sobel highlighted fine details and spatial frequencies, while fuzzy logic introduced a unique approach based on fuzzy conditions. Comparing the techniques, each displayed distinct strengths, with Canny being simple and robust, Sobel focusing on details, and fuzzy logic offering adaptability. The results suggest the crucial role of edge detection in forensic analysis, with potential future research on hybrid approaches for enhanced accuracy in real-world forensic cases.

Figure 8.2 shows the original images and their respective Canny contour identification results and Sobel contour identification results.

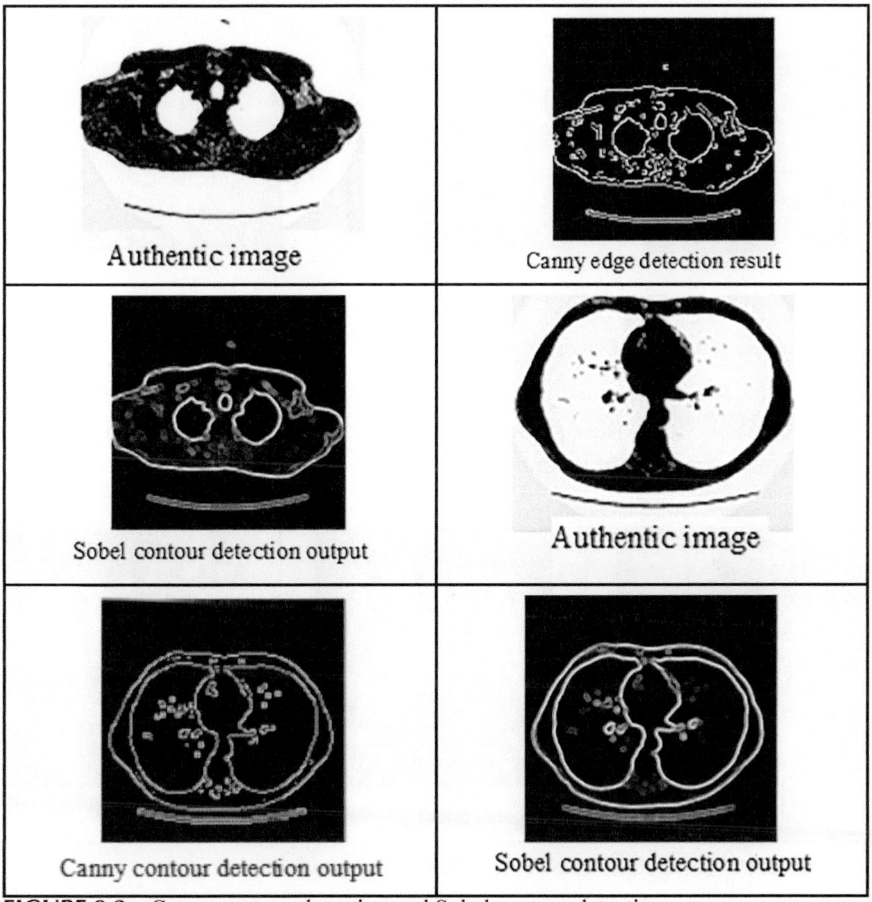

FIGURE 8.2 Canny contour detection and Sobel contour detection.

Figure 8.3 shows the original images and their respective fuzzy logic resultant outputs for organ size and shape to detect the disease.

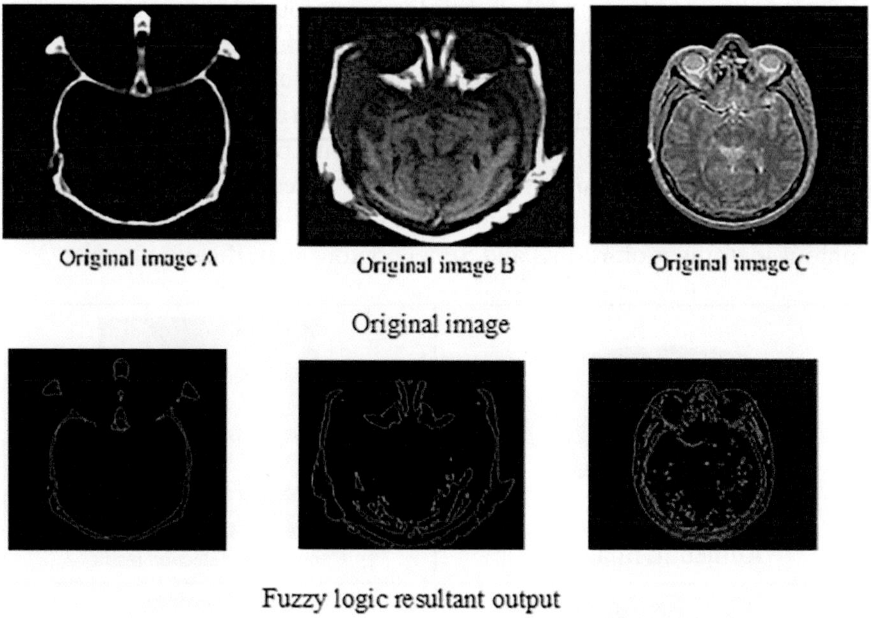

FIGURE 8.3 Fuzzy logic edge detection.

Figure 8.4 shows the original image (Author1) and its respective Sobel *X* and Sobel *Y* resultant outputs.

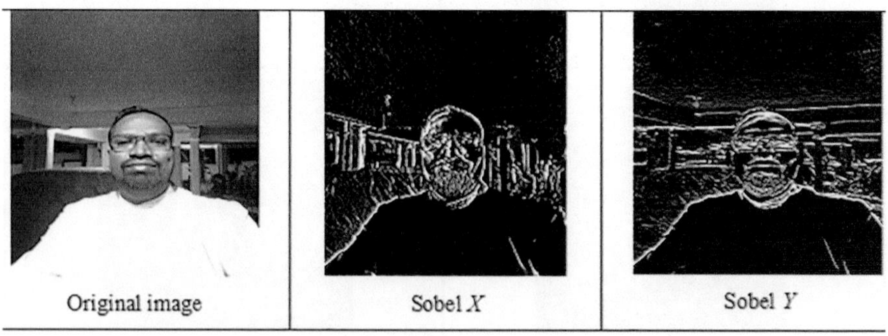

FIGURE 8.4 Sobel edge detection for forensic face recognition.

Figure 8.5 shows the source photograph (Author2) and its respective Canny resultant output.

Original photograph Canny border detection output

FIGURE 8.5 Canny border detection for forensic face recognition.

Figure 8.6 shows the original image and its respective fuzzy logic resultant output.

Original image Fuzzy logic edge detection resultant output

FIGURE 8.6 Fuzzy logic edge detection for forensic image recognition.

8.5 CONCLUSION AND FUTURE SCOPE

Based on the findings, the fuzzy logic algorithm performs well in detecting edges in medical and forensic images. It is effective in analyzing and identifying issues related to body parts, structures, and diseases. The fuzzy conditions it employs make it adaptable to various types of images. This aligns with the summary discussion, emphasizing that edge detection methods are essential in medical and forensic domains.

Looking ahead, further efforts can be directed toward improving the fuzzy logic algorithm for practical applications. Its effectiveness in diverse medical imaging scenarios can be explored further, along with integrating it with other techniques to enhance accuracy and flexibility in forensic image analysis.

DECLARATION

No AI tools or any other similar tools have been used in the content creation of this chapter.

KEYWORDS

- **canny border detection**
- **canny edge detection**
- **forensic image analysis**
- **fuzzy logic algorithm**
- **gradient operator**
- **image analysis accuracy**
- **MAX operator**
- **Sobel edge identification**

REFERENCES

Chinchkhede, D., & Uke, N. (2012). Image segmentation in video sequences using modified background subtraction. *International Journal of Computer Science & Information Technology (IJCSIT), 4*(1).

Das, S. (2015). Comparison of various edge detection technique. https://typeset.io/papers/comparison-of-various-edge-detection-technique-2yve640liw (accessed on 11th July 2025).

Dharampal, & Mutneja, V. (2015). Methods of image edge detection: A review. *Journal of Electrical & Electronic Systems, 4*, 150. https://doi.org/10.4172/2332-0796.1000150 (accessed on 11th July 2025).

Fisher, R. (2004). Website for Sobel edge detector. https://homepages.inf.ed.ac.uk/rbf/HIPR2/sobel.htm (accessed on 11th July 2025).

Haq, F. U. (2018). Website for online learning platform. https://www.educative.io/answers/what-is-canny-edge-detection (accessed on 11th July 2025).

Jain, S. (2008). Website for computer science education and skill development. *GeeksforGeeks*. https://www.geeksforgeeks.org/ (accessed on 11th July 2025).

Kumar, S., Upadhyay, A. K., Dubey, P., & Varshney, S. (2021). Comparative analysis for edge detection techniques. https://typeset.io/papers/comparative-analysis-for-edge-detection-techniques-20j3mm497q (accessed on 11th July 2025).

Meshram, S. A., & Lande, R. S. (2018). Traffic surveillance by using image processing. In *International Conference on Research in Intelligent and Computing in Engineering* (pp. 1–3). IEEE.

Prasetyo, A. B., Wahyudi, R., Tahyudin, I., Kusuma, S. F., Oktaviana, L. D., Barkah, A. S., & Artono, B. (2023). Comparative analysis of image on several edge detection. https://typeset.io/papers/comparative-analysis-of-image-on-several-edge-detection-vofbhg10 (accessed on 11th July 2025).

Srivastava, D., Kohli, R., & Gupta, S. (2017). Implementation and statistical comparison of different edge detection methods. https://typeset.io/papers/implementation-and-statistical-comparison-of-different-edge-142b44zxk1 (accessed on 11th July 2025).

Tagunde, G. A., Uke, N. J., & Banchhor, C. (2012). Detection, classification and recognition of road traffic signs using color and shape features. *International Journal of Advanced Technology & Engineering Research, 2*(4), 202–206.

Uke, N., & Thool, R. (2013). Moving vehicle detection for measuring traffic count using OpenCV. *Journal of Automation and Control Engineering, 1*(4), December.

Winarno, E., Hadikurniawati, W., Wibisono, S., & Septiarini, A. (2021). Edge detection method. https://typeset.io/papers/edge-detection-method-1p14os8t9r (accessed on 11th July 2025).

Withey, D. J., & Koles, Z. J. A review of medical image segmentation: Methods and available software.

CHAPTER 9

Enabling Communication Strategies: Indian Alpha-Sign Gesture Recognition

YOGESH BOROLE,* RAVINDRA PUNEKAR, and PRAVIN FUTANE

Department of Information Technology, BRACT's Vishwakarma Institute of Information Technology, Pune, Maharashtra, India

*Corresponding author

ABSTRACT

Effective communication, being a fundamental right, is often challenging for individuals with hearing impairments. The role of technology across various domains remains pivotal. Those deprived of natural senses require support through technological advancements. Natural gestures can facilitate communication for deaf or hearing-impaired individuals, bridging the gap with the general population.

This research addresses the imperative of empowering individuals with hearing impairments through Indian Sign Language (ISL) alphabet recognition (A–Z). The study introduces a methodology in which sign language gestures are provided as input to the system, utilizing a dataset comprising 26 ISL alphabets. A novel approach is introduced, employing convolutional neural networks (CNN), data processing, and one-hot encoding, achieving an exceptional accuracy rate of 99.33% in real-time ISL alphabet recognition from live video streams and subsequently generating predicted labels as text.

The CNN model, characterized by a carefully chosen architecture and optimized hyperparameters, demonstrates remarkable accuracy in ISL alphabet recognition. This chapter contributes significantly to the field by furnishing a robust and accurate ISL alphabet recognition tool, fostering the development of more inclusive and accessible communication platforms.

Intelligent Computing and Analytics in Image Processing. Samarjeet Borah, Jambi Ratna Raja Kumar, Laxmi Bewoor, Rahul Raghvendra Joshi, and Mahendra Prabhakar Deore (Eds.)

DOI: 10.1201/9781779640772-9

9.1 INTRODUCTION

Communication is a fundamental aspect of human interaction and an inherent human entitlement. It serves as the foundation upon which relationships, information exchange, and social participation are built. However, the journey toward effective communication is not universally smooth. Hearing impairment—whether congenital or acquired—can profoundly impact an individual's ability to perceive and understand auditory information. This disability not only affects verbal communication but also severs ties to ambient sounds that provide context and cues crucial for daily life. In India, where one-sixth of the world's population resides, hearing impairments are a significant concern. According to the National Sample Survey Office (NSSO) in India, approximately 6.3% of the Indian population, or around 63 million people, have some form of hearing impairment (Kaur and Kumar, 2017).

The need for technology-driven solutions becomes particularly compelling when considering the immense number of individuals who are Deaf or hard of hearing in India. Indian Sign Language (ISL) is an essential form of communication for this population. In a country as diverse as India, ISL represents not just a language but a cultural and linguistic bridge that connects millions. However, despite its significance, the effective recognition of ISL gestures through technology remains largely unexplored territory, holding immense potential to enhance communication strategies for individuals with hearing impairments.

Artificial intelligence (AI) and machine learning (ML) play a pivotal role in this context. Existing technologies have made strides in recognizing manual gestures but have often fallen short of achieving the high levels of accuracy and usability required for seamless communication. ML algorithms, such as convolutional neural networks (CNNs), offer the ability to recognize and interpret ISL gestures from live video streams with remarkable accuracy. This technology-driven approach holds the promise of revolutionizing the accessibility and inclusivity of ISL communication, allowing for more efficient and effective communication strategies.

The research utilizes a dataset of over 31,200 images representing ISL gestures, offering diverse examples of ISL alphabets, as shown in Figure 9.1. This study seeks to address this critical need by focusing on the recognition of ISL alphabets. Recognizing ISL alphabets not only enables the communication of individual letters but also serves as a foundation for constructing words, sentences, and conveying complex ideas. The ability to recognize ISL alphabets in real time through technology has the potential to revolutionize communication strategies for individuals with hearing impairments in India and beyond.

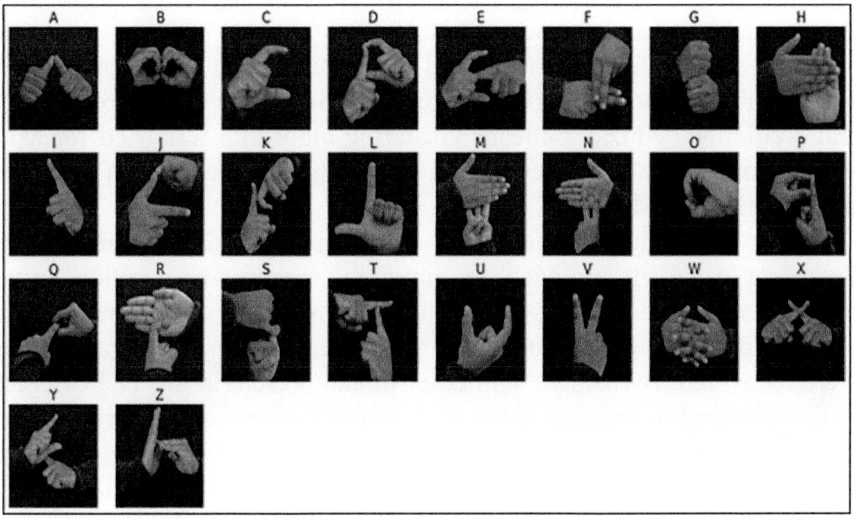

FIGURE 9.1 ISL dataset.

Source: Reprinted from Sabharwal et al., 2023. https://creativecommons.org/licenses/by/4.0/

9.2 LITERATURE SURVEY

Two techniques of classification algorithms are used by Rokade and Jadav (2017): neural network and support vector machine (SVM). Hu's moment is used for feature extraction. The segmentation phase is performed to detect the shape of the sign, and the detected region is transformed into a binary image. Artificial neural network (ANN) with 13 features provided an accuracy of 94.37%, while SVM with a polynomial kernel achieved an accuracy of 92.12%.

In the referenced study by Singh et al. (2015) use histogram of oriented gradients (HOG) for feature extraction. Techniques like K-nearest neighbors (KNN) and SVM were tested with geometric descriptors on a dataset consisting of a variable background. The HOG algorithm was used for modeling, and the KNN model yielded the best results with an accuracy rate of 88.46%. The results were further optimized using SVM with the HOG model, achieving an accuracy rate of 94.23%.

Rao et al. (2018) in the referenced study used a selfie-mode recording technique to create the dataset. About 200 signs from 5 different angles were taken under different environments. A CNN was utilized in the training process, employing varying sample sizes, each comprising multiple groups of individuals and different viewing perspectives. This strategy achieved an accuracy of 92.88%.

Paper by Muthu and Gomathi (2019) stated that the segmentation feature of OpenCV was performed for recognizing the signs. For training and prediction, a fuzzy c-means clustering algorithm was used for ML. A real-time system was developed, producing an accuracy of 75% for detecting ISL words.

Paper by Rekha, Bhattacharya, and Majumder (2011), the authors described a novel approach to automatic gesture recognition using multiclass nonlinear SVM. Using the principal curvature-based region (PCBR) detector, each hand's form, texture, and finger characteristics were recovered. By using a skin color model present in the YCbCr color space, the hand region was segmented. Experimental results indicated that the proposed system achieved an accuracy of 91.3% for dynamic gestures, with a recognition rate of 86.3%.

Paper by Kaur and Kumar (2017) explains the development of an ISL-based SMS generator for Deaf individuals. It introduces a graphical interface for sign sequence input, a language translator, and an SMS-to-speech translation system. The system's goal is to reduce communication barriers that exist between the hearing and Deaf communities, offering a platform for effective two-way communication. The system is designed to enable Deaf individuals to perform various communication tasks, including sending SMS messages, conveying urgent information, and interacting with visually impaired individuals through text.

In a study by Sawant and Kumbhar (2014), researchers created an accessible sign language recognition system utilizing MATLAB. This system can effectively identify 26 distinct gestures and is structured around four integral modules: preprocessing through hand segmentation, followed by feature extraction, sign recognition using principal component analysis (PCA), and converting recognized signs into both text and voice formats.

The research presented by Mangamuri, Jain, and Sharmay (2019) addresses the challenges of recognizing two-hand Indian sign language

(THISL) gestures. A comprehensive THISL dataset was developed, containing 26 gestures representing English alphabets, comprising 9100 images. Various ML classification models were applied, achieving an overall accuracy of 91.72%.

Using the ISL single-hand dataset, the researchers Shangeetha, Valliammai, and Padmavathi (2012) employed a simple technique for identifying movements. Based on hue, saturation, and intensity (HSI) segmentation, they detected fingertips and extracted information, including finger angles and states. They tested eight gestures and achieved an accuracy of 68.12%.

Paper by Kishore and Kumar (2012) presents a comprehensive video-based ISL recognition system (INSLR) that enhances interaction with hearing-impaired individuals. It integrates image processing, wavelet-based video segmentation, and fuzzy logic for gesture recognition, achieving a high recognition rate of 96%.

Paper by Singha and Das (2013) introduces a system utilizing weighted Euclidean distance with eigenvalue analysis for recognizing ISL. Skin filtering, cropping, feature extraction, and categorization are all part of the system. Using 240 images of 24 signs, a recognition rate of 97% was achieved.

Paper by Sahoo, Mishra, and Ahmed (2012) emphasize the need for ISL research and review international efforts in the UK, USA, Australia, Japan, China, and the Middle East. It highlights the development and history of British and American Sign Languages.

Paper by Sultana et al. (2018) discusses the difficulties that hearing-impaired and Deaf communities face in understanding Bangla Sign Language (BdSL). It introduces the "Concatenated BdSL Network," which achieves a 91.51% accuracy rate and highlights the importance of pose estimation in identifying BdSL symbols.

Paper by Jayadeep et al. (2020) encompasses Inception V3, long short-term memory (LSTM), and recurrent neural network (RNN) models for ISL recognition, using a dataset of 6000 sign words and 1100 videos, achieving an accuracy rate of 81%.

Bhavsar and Trivedi (2019) used a neuro-fuzzy (NF) classification algorithm to detect 10 commonly used ISL signs. Using skin color features, hand areas are identified, and hand shapes are separated. This method achieves an accuracy of 92% on the testing dataset (Table 9.1).

TABLE 9.1 Summary of Literature Survey

Paper References	Algorithm or Methodology	Dataset	Gesture Recognized	Accuracy
Rokade and Jadav (2017)	SVM, neural network	848 images (320×240 resolution)	17 different alphabets	SVM – 92.12% ANN – 94.37% with 13 features
Singh et al. (2015)	KNN, SVM	260 images from 7 different users	26 English alphabets	SVM – 94.23% KNN – 88.46%
Rao et al. (2018)	Multistage CNN	Self-created dataset	200 signs	92.88%
Muthu and Gomathi (2019)	Fuzzy C-means clustering	800 samples (words), 500 samples (sentences), 10 per sign	80 words and 50 sentences	75%
Rekha, Bhattacharya, and Majumder (2011)	Multiclass nonlinear SVM	Self-created dataset	30 ISL signs	Static – 91.3% Dynamic – 86.3%
Kaur and Kumar (2017)	SMS-based interface	250 sentences	Daily day-to-day sentences	–
Sawant and Kumbhar (2014)	Principal component analysis (PCA)	260 images	26 distinct gestures	Not mentioned
Mangamuri, Jain, and Sharmay (2019)	Machine learning algorithms	9100 images (50×50 resolution)	26 ISL alphabets	91.72%
Shangeetha, Valliammai, and Padmavathi (2012)	HSI-based segmentation	Not mentioned	8 ISL gestures	68.12%
Kishore and Kumar (2012)	ANN	800 images	80 different gestures	96%
Singha and Das (2013)	Eigenvalue-weighted Euclidean distance	240 images	24 signs	97%
Sahoo, Mishra, and Ahmed (2012)	Sign language study (various countries)	Not mentioned	Not mentioned	Not mentioned
Sultana et al. (2018)	CNN	11,061 still images	38 alphabets of Bangla Sign Language (BdSL)	91.51%
Jayadeep et al. (2020)	Inception V3, LSTM, RNN	6000 sign words, 1100 videos	ISL words	81%
Bhavsar and Trivedi (2019)	Neuro-fuzzy (NF) classification	100 images (10 per sign)	10 routine words	92%

9.3 EXPERIMENTAL METHODS

9.3.1 SOFTWARE TOOLS AND SETUP OR EXPERIMENTAL SETUP

In the experimental setup, visual studio code (VSCode), an adaptable integrated development environment (IDE), is used as the primary software tool. VSCode was chosen for its flexibility and support for various programming languages, making it ideal for implementing and testing the CNN model. To facilitate collaborative work, Jupyter Notebook (.ipynb) was employed using a relevant extension. Popular tools like TensorFlow and Keras were utilized for deep learning (DL), along with Python libraries such as MediaPipe, NumPy, OpenCV, and scikit-learn to work with data and evaluate the model.

9.3.2 DATASET

In this research, the standard Kaggle dataset is used to train the DL model for ISL recognition. This dataset is openly accessible under a Creative Commons license, ensuring broad availability for researchers. It comprises a total of 31,200 images, thoughtfully categorized into 26 distinct sign classes, specifically representing the 26 alphabet signs (Mangamuri, Jain, and Sharmay, 2019; Sawant and Kumbhar, 2014), from "A" to "Z." Within each class, there is a consistent count of 1200 images, each with dimensions of 128 × 128 pixels, and represented in the RGB (red, green, blue) format. The pixel values in this dataset range from 0 to 255 for each color channel. This dataset has been a crucial resource in the research, providing a strong foundation for the development and evaluation of ISL recognition models. The dataset images can be seen in Figure 9.1.

9.3.3 DATA PREPROCESSING

Data preprocessing is employed to decrease the computational resources needed and enhance the performance of the model. In this step, images and labels are first loaded from the training directory. For each label (representing sign classes), the image files are iterated through, resized to a consistent 64×64-pixel dimension, and then the images and their corresponding labels are collected into NumPy arrays.

9.3.4 DATA NORMALIZATION AND ONE-HOT ENCODING

Normalization is achieved by dividing RGB pixel values by 255.0 to scale them to the ideal [0, 1] range for effective neural network training. After normalization, one-hot encoding is applied to prepare the labels for DL. Since the dataset is categorical, regular ML algorithms may struggle with it. To make it easier for machines to interpret, one-hot encoding is used. This method represents categorical data as binary vectors, where each value is represented as 1 at its respective index, while all others are set to 0. This process transforms the label values into binary vectors, each representing a specific class, as shown in Figure 9.2. This is achieved using the "Keras to_categorical" function, which produces an array of one-hot encoded labels.

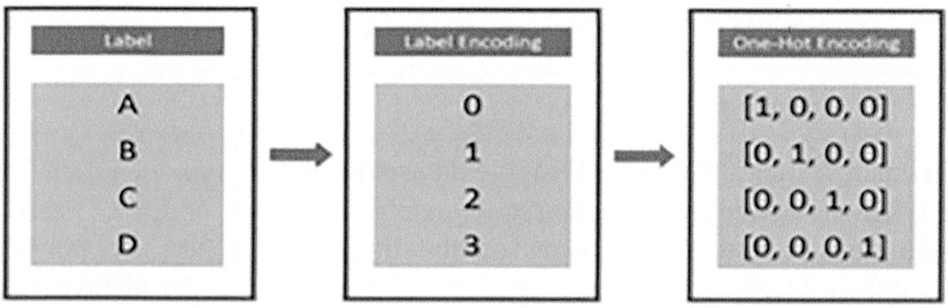

FIGURE 9.2 One-hot encoding.

9.3.5 CONVOLUTIONAL NEURAL NETWORK (CNN)

Convolutional neural networks (CNNs) have gained popularity as a technique for categorizing various types of tasks because they can automatically learn and extract important features from images. Artificial neurons, or nodes, in CNNs collect inputs, process them, and then provide outputs—similar to how neurons in the brain send information throughout the body (Figure 9.3).

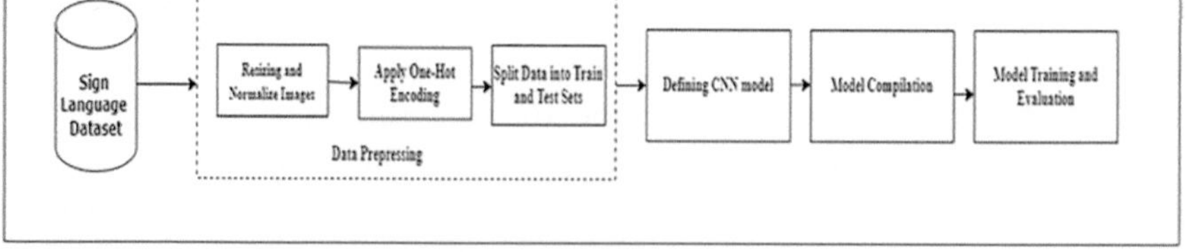

FIGURE 9.3 Block diagram for model training.

9.3.5.1 Model building

The proposed CNN architecture is specifically designed for processing 64×64-pixel images with three color channels, commonly known as RGB images. The architecture consists of a sequence of three convolutional layers, two max-pooling layers, five ReLU dropout layers, a final dense layer for classification, and one softmax function. Figure 9.4 shows the proposed CNN architecture.

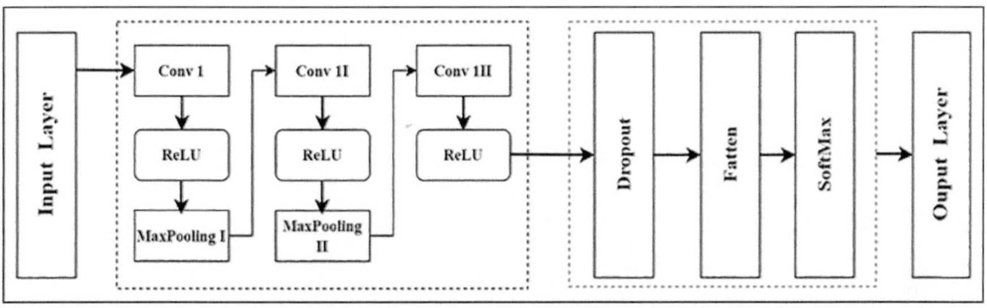

FIGURE 9.4 Diagram of the proposed CNN architecture.

Three Conv2D layers are used, each equipped with 64, 128, and 256 filters successively, with a 5×5 kernel size. A ReLU activation function is applied to introduce non-linearity; when the input is negative, the output is 0, as discussed below. The ReLU equation is given as (Figure 9.5):

$$f(z) = R(z) = \max(x, 0) \tag{1}$$

$$f(z) = R(z) = \begin{matrix} 0, \text{ for } z > 0 \\ z, \text{ for } z \leq 0 \end{matrix} \tag{2}$$

If the input value *z* is greater than 0 ($z > 0$), the output of the function *f(z)* is 0. If the input value *z* is less than or equal to 0 ($z \leq 0$), the output of the function *f(z)* is *z*. In other words, the function is linear for non-positive values and zero otherwise.

If the input value z is more than 0 (i.e., $z > 0$), the output of the function $f(z)$ is 0. This means that if z is positive, the output is set to 0. If the input value z is less than or equal to 0 (i.e., $z \leq 0$), the output of the function $f(z)$ is the same as the input value z. In other words, it remains linear for non-positive values of z.

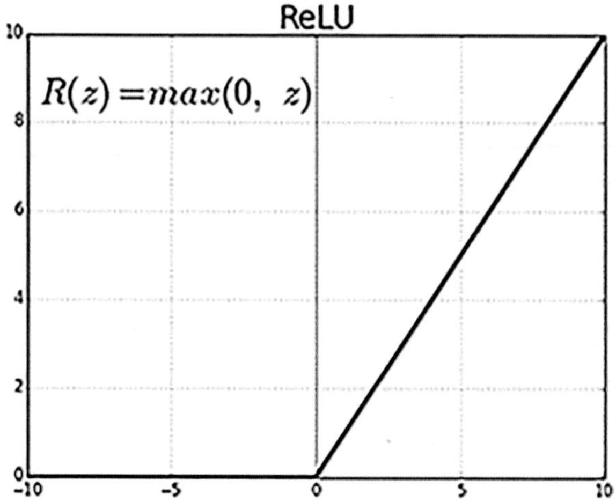

FIGURE 9.5 ReLU activation function.

To reduce the spatial dimensions of the input data, max pooling layers are used. These layers preserve essential features, enhance computational efficiency, and limit the model's susceptibility to overfitting. This is illustrated in Figure 9.6.

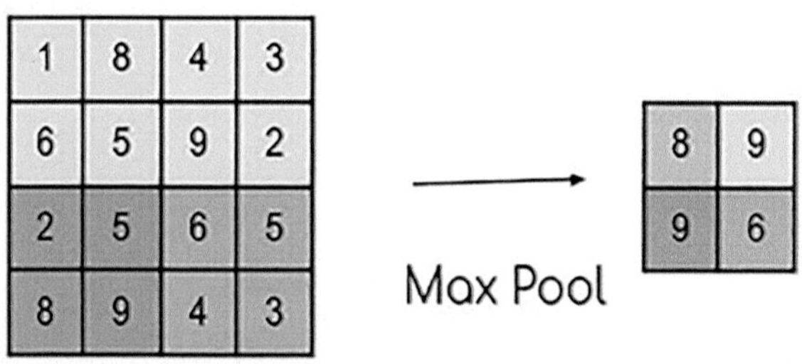

FIGURE 9.6 Max pooling example.

Dropout layers, with a dropout rate of 0.5, act as a form of regularization by randomly deactivating neurons during training. This helps prevent the model from overfitting.

At the end of the network, a flatten layer transforms the multi-dimensional convolutional output into a flat vector, as shown in Figure 9.8. The final dense layer, consisting of 26 neurons, uses the softmax activation function, as the task involves multi-class classification.

Using the softmax activation function, the neural network's raw outputs are transformed into a probability distribution over the input classes. As shown in Figure 9.7, this ensures that the sum of output probabilities is equal to 1, making it suitable for selecting the most likely class from a set of alternatives.

Figure 9.8 shows an example of a flatten (or straightening) layer that converts a two-dimensional (2D) data matrix into a one-dimensional vector, making it usable as input for a fully connected neural network classifier.

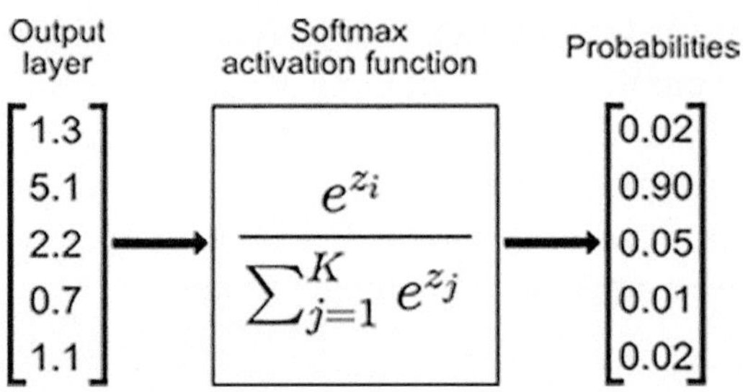

FIGURE 9.7 Equation of the softmax activation function.

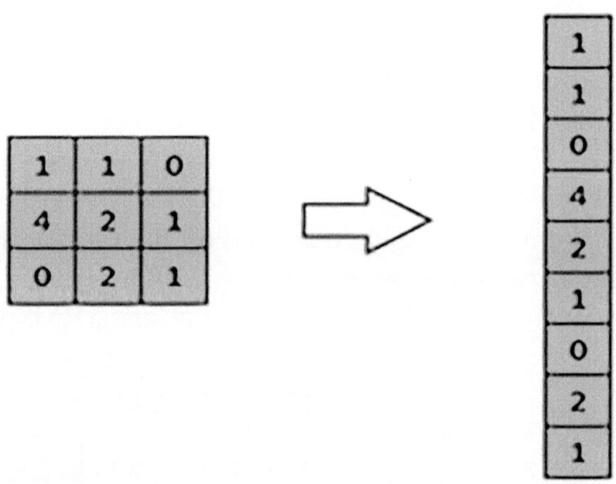

FIGURE 9.8 Flatten layer.

9.3.5.2 Model training

During the training phase, randomly selected images, as shown in Table 9.3, were taken from each alphabet set, associated with their respective labels, and fed into the CNN model. The learning rate was initially set to 0.001.

The model was trained with various hyperparameters, including learning rate, batch size, image filtering techniques, number of training epochs, and other parameters, as described in Table 9.2.

Two crucial callback functions, ReduceLROnPlateau and EarlyStopping, have been implemented to enhance the training process. The ReduceL-ROnPlateau callback is designed to dynamically adjust the learning rate. It continuously monitors the training loss (monitor='loss') and reduces the learning rate after two consecutive epochs (patience=2) if no progress is observed. The cooldown period is set to 0, meaning there is no delay before modifying the learning rate. The EarlyStopping callback ensures efficient training by monitoring the training accuracy. It requires a minimum improvement of 0.0001 (min_delta=1e-4) and terminates training if no such improvement is detected over two consecutive epochs. By combining these callback methods, overfitting and inefficiencies in the training process are avoided, helping ensure successful model training.

TABLE 9.2 Hyperparameters for Overseeing the Training Procedure

Epochs	Patience	Monitor (ReduceLROnPlateau)	Cooldown	min_delta	Monitor (EarlyStopping)
10	2	loss	0	1e-4	Accuracy

To initiate model training, the first step is to compile the CNN model, as described earlier. The model was compiled using accuracy as the evaluation metric, RMSprop as the optimizer, and the loss function as categorical cross-entropy. The model was then trained on the training dataset for 10 epochs using the fit method. See the prototype of the system in Figure 9.3.

Mathematical equations used in model training for the RMSprop optimizer:

$$Accuracy = \frac{TP + TN}{TP + TN + FP + FN} \tag{3}$$

where; $E[g^2]_t$ is the moving average of squared gradients; g^2_t is the square of the gradient at time step t; η is the learning rate; and ε is a small constant added for numerical stability.

9.4 REAL-TIME SIGN PREDICTIONS

After completing the training of the CNN model on the defined dataset, the model is saved as Mymodel.h for real-time predictions. OpenCV is used to access the live camera feed. Within this setup, a custom hand detection module using MediaPipe enables the detection and tracking of hand landmarks within the video stream.

To recognize hand motions used as signs, each video frame is continuously recorded and analyzed. A region of interest (ROI) is defined within each frame to isolate the hand gesture. This isolated region is then resized and normalized to a consistent input size for the CNN model. The model predicts the sign being made in the ROI and returns the class label with the highest predicted probability. This label corresponds to the recognized sign. The recognized sign is displayed in real time on the video frame using OpenCV (*see* Figure 9.9).

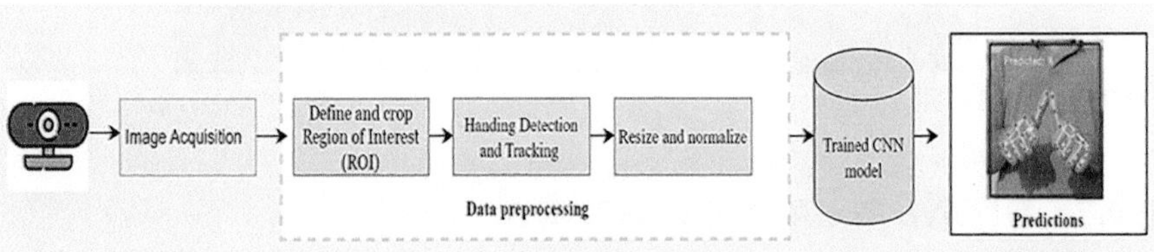

FIGURE 9.9 Flow diagram of ISL model real-time testing.

9.4.1 *IMAGE CAPTURING*

For image capturing, the OpenCV library was used to connect to the computer's built-in camera, allowing real-time video to be captured for experimentation. The experiment was carried out on an Intel i3-core system with a 4MP camera setup. OpenCV is a widely used tool for simplifying camera communication and handling video frames, making it valuable for various applications.

9.4.2 *HAND DETECTION*

In this research, a custom Python class, handDetector, is implemented. It enables hand detection and landmark tracking in real time by using the MediaPipe library. The handDetector class includes several adjustable parameters, offering flexibility in detecting and tracking hands. These parameters include tracking modes, the number of hands to detect, model complexity, confidence thresholds, and more.

Once a hand is detected, the function converts the frame to the required RGB format and draws landmarks on the hand in both still and moving video frames, as shown in Figure 9.10.

FIGURE 9.10 Hand detection.

9.4.3 *DEFINE REGION OF INTEREST (ROI)*

In this step, the concept of a ROI is used as a critical component of hand gesture recognition. The ROI is determined by defining the top-left (x_1, y_1) and bottom-right (x_2, y_2) corner points, creating a square-shaped area within the frame. Mathematically, the ROI can be defined as:

$$\text{Cropped Frame} = \text{Frame}\ (x_1{:}x_2, y_1{:}y_2) \tag{4}$$

Where *Frame* represents the video frame captured through the live camera, and *Cropped Frame* refers to the isolated region, as shown in Figure 9.11. This technique increases the accuracy and efficiency of hand gesture recognition.

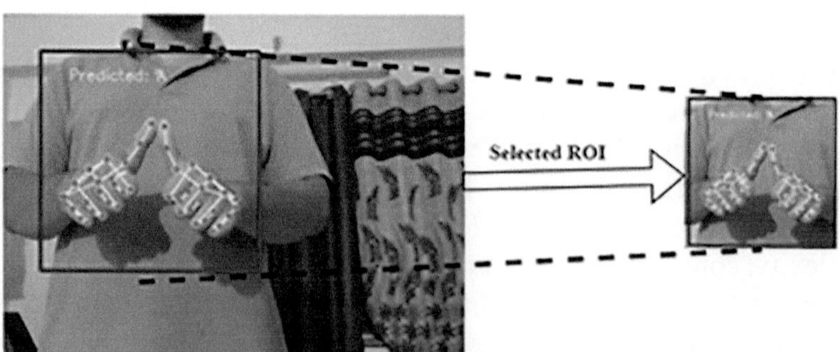

FIGURE 9.11 Region of interest (ROI) for hand gesture recognition.

9.4.4 *FRAME PROCESSING*

After cropping to the ROI, uniformity is achieved by resizing the images to a fixed dimension of 64×64 pixels. The pixel values are then normalized by dividing them by 255, scaling them to the range [0, 1].

9.4.5 *SIGN PREDICTION*

After preprocessing, the image is passed through the trained model, where the classifier determines the corresponding label. The system then checks a dictionary using this label as a key, and the dictionary returns the relevant sign as the value. The predicted label is then displayed on the screen in real time. The output of some real-time predictions is shown in Figure 9.12.

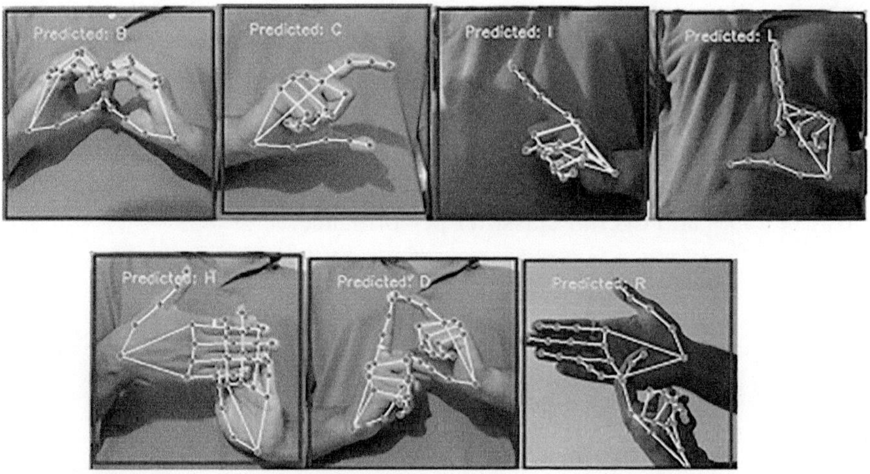

FIGURE 9.12 Real-time prediction results (snapshot of the trained model).

9.5 RESULTS AND DISCUSSION

The model's performance is tested using a combination of VSCode and Google Colaboratory. As previously discussed, a dataset comprising 26 distinct sign language classes was selected, totaling 31,200 images. To rigorously evaluate the model, the dataset is divided using four different split ratios: 90:10, 80:20, 70:30, and 60:40 for training and validation, respectively.

This study employed the RMSprop optimizer, a batch size of 128, and 10 training epochs. The experimental model's performance is summarized in Table 9.3.

TABLE 9.3 Model Performance on Different Dataset Splits

Dataset Split	Accuracy
90:10	0.9931
80:20	0.9929
70:30	0.9933
60:40	0.9920

In Table 9.3, it can be observed that the model's performance remains consistently high across varying dataset splits, with accuracy scores ranging from 99.20% to 99.33%. Among the splits, the 70:30 arrangement exhibited the highest accuracy of 99.33%, demonstrating the model's robustness and efficiency in recognizing sign language gestures.

Figure 9.13 displays the loss graph, while Figure 9.14 shows the accuracy graph of the experiment. Precision, recall, and F1-score values obtained during testing are also shown in Table 9.3. These metrics are valuable for assessing the model's effectiveness in correctly identifying positive cases and overall classifier performance.

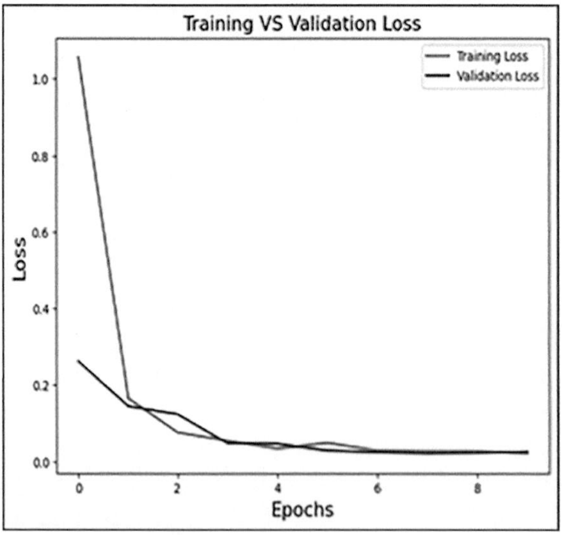

FIGURE 9.13 Train and validation loss.

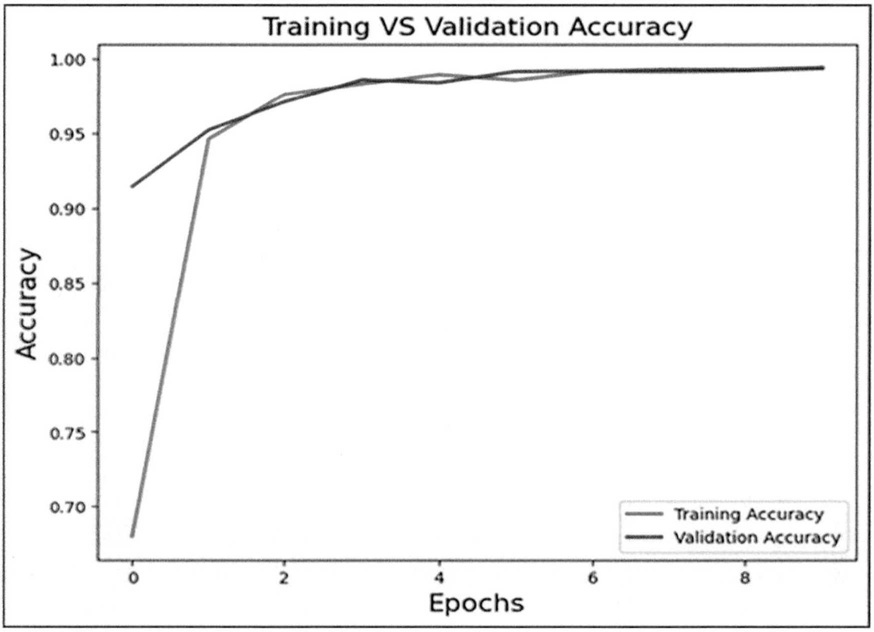

FIGURE 9.14 Train and validation accuracy.

9.5.1 QUANTITATIVE ANALYSIS

Figure 9.15 displays the F1-score, recall, and precision values obtained during the testing phase. The metrics are based on the terms true positive (TP), false positive (FP), true negative (TN), and false negative (FN). The formulas used to calculate these metrics are provided below:

$$Accuracy = \frac{TP + TN}{TP + TN + FP + FN} \tag{5}$$

$$Precision = \frac{TP}{TP + FP} \tag{6}$$

Precision assesses the capability of the classifier to accurately identify TP samples as positive.

$$Precision = \frac{TP}{TP + FN} \tag{7}$$

Recall, on the other hand, quantifies the fraction of actual positive samples that have been correctly identified as positive.

$$F1-score = \frac{2\times\text{Precision}\times\text{Recall}}{\text{Precision}\times\text{Recall}} \quad (8)$$

The F1-score is a combined metric that balances both precision and recall. It is particularly useful for evaluating the overall performance of the model, especially when dealing with imbalanced classes.

Class	Label	Precision	Recall	F1-Score
A	0	1.00	1.00	1.00
B	1	1.00	1.00	1.00
C	2	1.00	1.00	1.00
D	3	1.00	1.00	1.00
E	4	1.00	1.00	1.00
F	5	1.00	1.00	0.99
G	6	1.00	1.00	1.00
H	7	1.00	1.00	1.00
I	8	1.00	1.00	1.00
J	9	1.00	1.00	1.00
K	10	1.00	1.00	1.00
L	11	1.00	1.00	1.00
M	12	0.99	1.00	1.00
N	13	1.00	1.00	1.00
O	14	1.00	1.00	1.00
P	15	1.00	1.00	1.00
Q	16	1.00	1.00	1.00
R	17	1.00	1.00	1.00
S	18	1.00	1.00	1.00
T	19	1.00	0.98	1.00
U	20	1.00	1.00	1.00
V	21	0.99	0.99	1.00
W	22	1.00	1.00	1.00
X	23	1.00	1.00	1.00
Y	24	1.00	1.00	1.00
Z	25	1.00	0.99	1.00

FIGURE 9.15 Classification report: Precision, recall, and F1-score.

9.5.2 COMPARISON WITH EXISTING SYSTEMS

Table 9.4 shows a comparison of various ML algorithms and highlights how differently they perform under varying conditions when used for the recognition of ISL. These factors include dataset size, method, technique used, and accuracy. In this study, various algorithms are considered, such as SVM, LSTM, ANN, Inception V3, clustering techniques, and others. The aforementioned factors are used for comparison and evaluation of the models.

In a study by Rekha, Bhattacharya, and Majumder (2011), using a dataset of 1000 samples, the SVM technique with PCBR features achieved an accuracy of 91.3%. However, in the case of dynamic gestures, the accuracy marginally declined, as PCBR features tend to be less stable. Meanwhile, the Fuzzy C-Means Clustering method used by Muthu and Gomathi (2019) with a dataset of 1,200 images produced an accuracy of 75%. This suggests that fuzzy clustering may not be ideal for sign language recognition, especially when features exhibit high similarity or lack clear boundaries between sign movements. In such cases, the method may result in less effective and precise recognition, placing it behind other techniques discussed.

In the case of Inception V3 combined with LSTM (Jayadeep et al., 2020), a larger dataset of nearly 6000 images was used for training. Surprisingly, this model achieved only 81% accuracy, implying that the complexity of the model's architecture may have impacted its performance. By contrast, PCA with fuzzy inference (Kishore and Kumar, 2012), using only 800 sign gesture images, achieved an exceptionally high accuracy of 96%. Elliptical Fourier descriptors and discrete wavelet transform (DWT) were used in that system to extract features, chosen for their proven effectiveness in capturing the unique characteristics of sign gestures and improving recognition capability.

The proposed CNN model, trained on a significantly larger dataset of 31,200 ISL sign alphabets, ranks highest with an unmatched accuracy of 99.33% compared to similar studies. The large dataset and carefully selected hyperparameters contribute to the outstanding performance of the proposed model, distinguishing it from other existing systems in terms of accuracy and reliability.

TABLE 9.4 Comparative Performance Analysis

Paper References	Algorithm or Methodology	Dataset	Accuracy
Rekha, Bhattacharya, and Majumder (2011)	SVM	Up to 1000 samples	91.3%
Jayadeep et al. (2020)	Inception V3, LSTM	6000 sign words	81%
Kishore and Kumar (2012)	PCA with fuzzy inference	800 images	96%
Muthu and Gomathi (2019)	Fuzzy C-means clustering	1200 images	75%
Proposed method	CNN	31,200 images of ISL alphabets	99.33%

9.6 CONCLUSION

This research aimed to address the critical need for empowering individuals with hearing impairments through ISL alphabet recognition. Using an innovative approach that incorporates CNNs and a dataset comprising 31,200 images representing 26 ISL alphabets, the model achieved an exceptional real-time accuracy rate of 99.33% in recognizing ISL gestures from live video streams.

The study demonstrates significant potential to improve communication for Deaf and hard-of-hearing communities in India and beyond. Compared to other models, the proposed CNN-based system outperforms previous methods, providing a more effective solution for inclusive communication.

Additionally, the system's accuracy remained consistent across different dataset splits, indicating robustness and reliability in diverse scenarios. The model's ability to recognize signs in real time using a live video feed further emphasizes its practical applicability. This work represents a meaningful step toward making communication more inclusive and accessible for individuals with hearing impairments.

9.7 FUTURE WORK

The proposed method has produced positive results, but there is still scope for improvement. The current system is limited to recognizing only the ISL alphabet, from A to Z. Future research could focus on making the model more robust by expanding and increasing the vocabulary of ISL, enabling it to recognize a larger set of gestures beyond the alphabet.

Additionally, exploring hybrid models that can translate signs into speech in real time—not just text—would enhance the system's applicability. Future work could also include the development of a user-friendly interface and the integration of the model into various applications, devices, and services. This would provide greater independence and convenience for deaf individuals, making the system more accessible in everyday use.

The ultimate goal is to build a future where communication is truly inclusive and limitless, ensuring that every individual's voice is heard.

DECLARATION

No AI tools or any other similar tools have been used in the content creation of this chapter.

KEYWORDS

- **artificial neural network**
- **convolutional neural network**
- **gesture recognition**
- **Indian sign language**
- **long short-term memory**
- **principal curvature based region**
- **support vector machine**
- **TensorFlow**
- **region of interest**

REFERENCES

Bhavsar, H., & Trivedi, J. (2019). Hand gesture recognition for Indian Sign Language using skin color detection and correlation-coefficient algorithm with neuro-fuzzy approach. In *2019 International Conference on Advances in Computing, Communication and Control (ICAC3)*, Mumbai, India (pp. 1–5). https://doi.org/10.1109/ICAC347590.2019.9036832 (accessed on 11th July 2025).

Jayadeep, G., V, V. N., Venugopal, V., S, V., & M, G. (2020). Mudra: Convolutional neural network based Indian Sign Language translator for banks. In *2020 4th International Conference on Intelligent Computing and Control Systems (ICICCS)*. https://doi.org/10.1109/ICICCS48265.2020.9121 (accessed on 11th July 2025).

Kaur, R., & Kumar, P. (2017). Sign language-based SMS generator for hearing impaired people. In *2017 International Conference on Computational Intelligence in Data Science (ICCIDS)*, Chennai, India (pp. 1–5). https://doi.org/10.1109/ICCIDS.2017.8272655 (accessed on 11th July 2025).

Kishore, P., & Kumar, P. (2012). A video based Indian Sign Language recognition system (INSLR) using wavelet transform and fuzzy logic. *IACSIT International Journal of Engineering and Technology, 4*(5), 537–542. http://www.ijetch.org/papers/427-C074.pdf (accessed on 11th July 2025).

Mangamuri, L. S. T., Jain, L., & Sharmay, A. (2019). Two hand Indian Sign Language dataset for benchmarking classification models of machine learning. In *2019 International Conference on Issues and Challenges in Intelligent Computing Techniques (ICICT)*, Ghaziabad, India (pp. 1–5). https://doi.org/10.1109/ICICT46931.2019.8977713 (accessed on 11th July 2025).

Muthu Mariappan, H., & Gomathi, V. (2019). Real-time recognition of Indian Sign Language. In *2019 International Conference on Computational Intelligence in Data Science*

(ICCIDS), Chennai, India (pp. 1–6). https://doi.org/10.1109/ICCIDS.2019.8862125 (accessed on 11th July 2025).

Rao, G. A., Syamala, K., Kishore, P. V. V., & Sastry, A. S. C. S. (2018). Deep convolutional neural networks for sign language recognition. In *2018 Conference on Signal Processing and Communication Engineering Systems (SPACES)*, Vijayawada, India (pp. 194–197). https://doi.org/10.1109/SPACES.2018.8316344 (accessed on 11th July 2025).

Rekha, J., Bhattacharya, J., & Majumder, S. (2011). Shape, texture and local movement hand gesture features for Indian Sign Language recognition. In *3rd International Conference on Trendz in Information Sciences & Computing (TISC 2011)*, Chennai, India (pp. 30–35). https://doi.org/10.1109/TISC.2011.6169079 (accessed on 11th July 2025).

Rokade, Y., & Jadav, P. (2017). Indian sign language recognition system. *International Journal of Engineering and Technology, 9*, 189–196. https://doi.org/10.21817/ijet/2017/v9i3/170903S030 (accessed on 11th July 2025).

Sabharwal, S., Singla, P. (2023). Alphabet-Level Indian Sign Language Translation to Text Using Hybrid-AO Thresholding with CNN. Intelligent Automation & Soft Computing, 37(3), 2567–2582. https://doi.org/10.32604/iasc.2023.035497

Sahoo, A. K., Mishra, G., & Ahmed, P. (2012). A proposed framework for Indian Sign Language recognition. *International Journal of Computer Applications (IJCA), 5*(2), 158–169.

Sawant, S. N., & Kumbhar, M. S. (2014). Real-time sign language recognition using PCA. In *2014 IEEE International Conference on Advanced Communications, Control and Computing Technologies*, Ramanathapuram, India (pp. 1412–1415). https://doi.org/10.1109/ICACCCT.2014.7019333 (accessed on 11th July 2025).

Shangeetha, R., Valliammai, V., & Padmavathi, S. (2012). Computer vision-based approach for Indian Sign Language character recognition. In *2012 International Conference on Machine Vision and Image Processing (MVIP)*. https://doi.org/10.1109/MVIP.2012.6428790 (accessed on 11th July 2025).

Singh, A., Arora, S., Shukla, P., & Mittal, A. (2015). Indian Sign Language gesture classification as single or double handed gestures. In *2015 Third International Conference on Image Information Processing (ICIIP)*, Waknaghat, India (pp. 378–381). https://doi.org/10.1109/ICIIP.2015.7414800 (accessed on 11th July 2025).

Singha, J., & Das, K. (2013). Indian Sign Language recognition using eigenvalue weighted Euclidean distance-based classification technique. *International Journal of Advanced Computer Science and Applications (IJACSA), 4*(2). https://doi.org/10.14569/IJACSA.2013.040228 (accessed on 11th July 2025).

Sultana, S., et al. (2018). Bangla Sign Language detection using SIFT and CNN. In *2018 9th International Conference on Computing, Communication and Networking Technologies (ICCCNT)* (pp. 1–6). https://doi.org/10.1109/ICCCNT.2018.8493976 (accessed on 11th July 2025).

CHAPTER 10

Evaluating Face Detection for Real-Time Emotion Recognition

HARSHVARDHAN PATIL,* PAWAN WAWAGE, and YOGESH DESHPANDE

Department of Information Technology, BRACT's Vishwakarma Institute of Information Technology, Pune, Maharashtra, India

*Corresponding author

ABSTRACT

Facial emotion recognition (FER) has been a rapidly advancing field, with various methods and techniques proposed to address this challenge. One of the most widely used approaches involves leveraging facial expressions captured in images. In this paper, we investigate the impact of applying face detection to test images and analyze the differences in results generated by a convolutional neural network (CNN) model. In this work, we trained a CNN model on the FER2013 dataset to classify three basic emotions: happy, sad, and neutral. The model achieved a training accuracy of 83% and a validation accuracy of approximately 79%. Samples from the AffectNet dataset are used as the unseen test data for the model. Face detection is applied to the test images to isolate the relevant facial regions that are most useful for emotion prediction.

10.1 INTRODUCTION

Recognizing emotions has been a subject of extensive study, and real-time emotion recognition holds considerable significance in contemporary applications (Haghpanah et al., 2022). Human emotions, while diverse,

Intelligent Computing and Analytics in Image Processing. Samarjeet Borah, Jambi Ratna Raja Kumar, Laxmi Bewoor, Rahul Raghvendra Joshi, and Mahendra Prabhakar Deore (Eds.)

DOI: 10.1201/9781779640772-10

include certain fundamental expressions that are universally consistent (Ekman and Friesen, 1971). Numerous studies have endeavored to address this challenge by devising novel and enhanced methodologies, leading to the creation of various computer systems equipped with underlying models designed to learn and recognize emotions.

Several existing systems employ a variety of models and algorithms. For instance, Reyes et al. utilized the SVM algorithm for classifying different emotions showcased by humans. However, convolutional neural networks (CNNs) have demonstrated exceptional capabilities in feature extraction and efficient computations (Krizhevsky, Sutskever, and Hinton, 2017). Recently proposed methods predominantly leverage CNNs for emotion detection, yielding noteworthy results and establishing them as widely adopted models for emotion recognition (Gowri, Rafeeq, and Devipriya, 2021; Jain, Shamsolmoali, and Sehdev, 2019; Khaireddin and Chen, 2021; Mehendale, 2020; Pitaloka et al., 2017; Pranav et al., 2020; Tümen, Söylemez, and Ergen, 2017).

Despite the multifaceted nature of emotion recognition as a problem, the choice of datasets depends on the specific solution being implemented. Among the myriad proposed solutions, facial expression image datasets, particularly the standardized FER2013 dataset, have consistently featured. The human accuracy on the FER2013 dataset is estimated to be 65.5% (Goodfellow et al., 2013).

In addition to enhancing the performance of underlying ML or deep learning (DL) models, the quality of input data is a critical factor influencing system performance during testing on unseen or real-time data. Preprocessing the input data significantly impacts model performance (Pitaloka et al., 2017). Ongoing developments in computer vision strive to enhance classification accuracy for such challenges (Jain, Shamsolmoali, and Sehdev, 2019; Pitaloka et al., 2017; Tümen, Söylemez, and Ergen, 2017).

This work aims to investigate the impact of applying a face detection algorithm to extract the region of interest (ROI) before feeding it to the model, compared to the model's performance when no such techniques are applied and raw data is used. The study also evaluates the suitability of different face detection algorithms for real-time face detection systems, considering factors such as time overhead, precision, and accuracy.

10.2 RELATED WORKS

Hagpanah et al. (2022) proposed a lightweight algorithm for feature extraction from static images that were used to train the MLP neural network model. The authors also applied preprocessing, landmark detection, and feature extraction from images by dividing them into different sections. Based on this, they extracted geometric- and texture-based features and created a normalized set of features, which was then fed to the 3-layered MLP for training. The proposed system performed with an accuracy of 96% on test data.

Khaireddin et al. (2021) claimed to achieve the highest single-network classification accuracy of 73.28% on the FER2013 dataset. They adopted a model based on the VGGNet architecture and rigorously trained its hyperparameters along with optimization techniques, learning rate schedulers, and weight decomposers. The authors also applied data augmentation techniques to account for variability in the dataset.

Pitaloka et al. (2017) emphasized better preprocessing techniques of the available dataset to enable effective extraction of features from the provided images, such as face detection, cropping, resizing, addition of noise (data augmentation), and data normalization. Raw, non-preprocessed data exhibited an average accuracy of 61.81%, whereas applying these techniques improved performance and achieved a maximum accuracy of 93.14%.

Reyes et al. (2020) implemented the HAAR-CASCADE algorithm to detect faces (regions of interest). The authors used feature descriptors—specifically, the histogram of oriented gradient (HOG) and the local binary pattern (LBP)—to capture relevant information from facial images. A ML algorithm (SVM) was trained on this information and implemented in the NAO robot.

Gowri et al. (2021), in their paper, showcased the importance of facial feature extraction and face detection. The Dlib facial landmark and the Harris Corner feature identification method were used, respectively, to effectively identify and extract features from facial images, allowing for precise recognition and extraction of key facial attributes. Following the process of feature extraction, the extracted facial features underwent preprocessing through various filters, ensuring optimal refinement before being fed into the classifier for final output. Furthermore, data augmentation was used to address overfitting problems. The FER2013 dataset was utilized, and the model performed with an accuracy of 73%.

Pranav et al. (2020) proposed a CNN model that classifies five different emotions. The authors utilized a manually collected image dataset, which was used to train, test, and validate the model. They claimed to achieve an accuracy of 78.04%.

In all the systems referred to so far, it is evident that the data preprocessing stage plays a crucial role in enhancing the feature extraction capabilities of the overall system. It encourages the model to capture and learn from the most relevant and important features, which significantly impacts the performance of the overall system, thereby signifying the importance of preprocessing steps in the overall implementation of these systems.

10.3 DATASETS USED

10.3.1 FER2013

Facial emotion recognition (FER) is a well-known standard dataset introduced in 2013 at the ICML conference. It contains images of individuals expressing various emotions, collected from various sources on the internet. This dataset contains images classified into seven distinct emotions: disgust, surprise, happiness, sadness, neutrality, anger, and fear. Each image is a grayscale image with a size of 48×48 pixels. The distribution of sample images across each emotion class is listed below (Table 10.1).

TABLE 10.1 FER2013 Dataset

Emotion	Number of Samples
Anger	4953
Disgust	547
Fear	5121
Happiness	8989
Sadness	6077
Surprise	4002
Neutral	6198

As evident from the above table, the image samples across each class of emotion are unevenly distributed, with the highest number of samples being 8989 for the emotion happiness, and the lowest number of samples being 547 for the emotion disgust.

10.3.2 AFFECTNET

AffectNet is another dataset specifically designed for the purpose of FER. It comprises over 1 million facial images collected from various sources on the internet and categorized into different emotions. The dataset includes images representing a wide range of ages, ethnicities, and lighting conditions, rendering it suitable for training and testing emotion recognition models in real-world scenarios. However, for this study, we utilized a subset of the AffectNet dataset, which contains samples for eight different emotions: anger, happiness, contempt, sadness, disgust, fear, surprise, and neutrality. Each image is a color (3-channel R, G, B) image with a size of 96×96 pixels.

10.4 IMPLEMENTATIONS

10.4.1 DATASET, PREPROCESSING, AND AUGMENTATION

The FER2013 dataset, a standard dataset, served as the foundation for training the underlying CNN model. Due to an uneven distribution of samples among different classes, only the top three classes—happiness, sadness, and neutrality—were retained for training, while all other classes were discarded. Each sample in the dataset is a single-channel grayscale image with dimensions of 48×48 pixels.

In the pursuit of enhancing data diversity and bolstering the model's resilience, data augmentation techniques were employed. These techniques encompassed operations such as image rotation, shifting, zooming, and horizontal flipping applied to randomly selected images. Data augmentation is instrumental in enabling the model to generalize effectively and perform optimally across a spectrum of real-world scenarios, accommodating slight variations in the images.

Furthermore, the dataset was divided into two subsets: a training set and a validation set in a 90:10 split ratio. This division is essential for training the model on one subset while validating its performance on another, ensuring the model's generalizability and assessing its effectiveness in recognizing emotions across a wide range of cases.

10.4.2 TRAINING AND INFERENCE

The training phase encompassed 100 epochs, employing a batch size of 32 samples and optimizing the categorical cross-entropy loss. The N-Adam optimizer was selected, with a learning rate step size set to 0.001. Additionally, two callback methods were implemented to mitigate the risk of overfitting. The first involved early stopping of training epochs, and the second dynamically reduced the learning rate based on observed conditions in the monitored parameter, which, in this instance, was the validation accuracy.

10.4.3 CNN ARCHITECTURE

The implemented model utilized a sequential CNN, as illustrated in Figure 10.1. The preprocessed images from the dataset, each with dimensions of 48×48×1, underwent feature extraction within the convolutional network. This involved the application of three sets of convolutional layers, each comprising two convolution layers, batch normalization, and a single max-pooling and dropout layer.

The filters/kernels embedded in the convolution layers play a crucial role in feature extraction, detecting sets of valuable patterns and features. Normalization layers serve to normalize activations from preceding layers, thereby facilitating faster calculations, improved convergence, and enhanced generalization. The max-pooling layer employs a pool size of 2×2, contributing to the extraction of the most significant features (those with maximum values). This process results in the discarding of less significant features, consequently reducing the overall size of the feature map. This reduction in computational complexity serves to control overfitting. The features extracted by the convolution layers manifest as a 2D array, given that the images are in grayscale format. Consequently, it becomes necessary to convert this representation into a singular 1D array. This conversion allows for feeding the extracted features to the fully connected dense layers, which can identify the correct weighted relationships among the most pivotal features obtained from the input image. In the final dense layer, the softmax function is applied. This choice is pertinent due to the nature of emotion detection being a multi-class classification problem, with classes denoted as 0: "Happy," 1: "Sad," and 2: "Neutral."

10.4.4 EMOTION PREDICTION

In consideration of the real-time application scope of this study, the AffectNet dataset was selected to assess the model's performance.

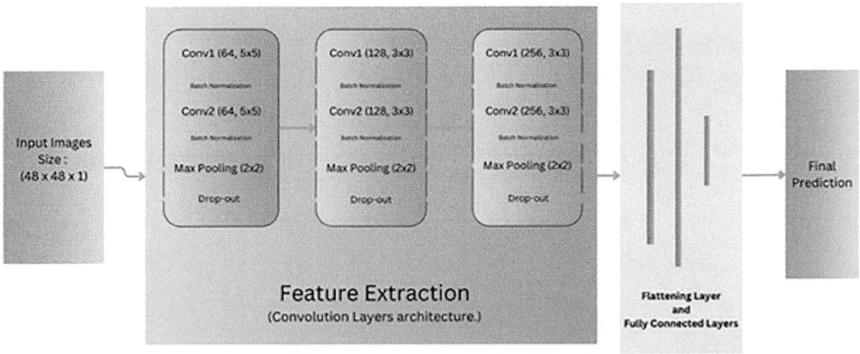

FIGURE 10.1 CNN model architecture.

The images within the AffectNet dataset are color images, each measuring 96×96 pixels.

Prior to input into the trained CNN model, each image underwent a series of preprocessing stages. These stages included face detection and subsequent cropping of the image frame based on the detected facial region, conversion of the cropped image from color to grayscale, and resizing the image to dimensions of 48×48 pixels. The conversion to grayscale and resizing to 48×48 pixels are integral steps in the preprocessing pipeline. These procedures play a crucial role in reducing computational complexity, thereby expediting computations—a pivotal requirement for real-time applications.

Subsequently, the processed image frames were introduced into the CNN model, which, leveraging the features acquired during its training phase, generated predictions based on the input data.

10.5 EXPERIMENTS AND RESULTS

The CNN model trained on the FER2013 dataset exhibited a training accuracy of around 83% and a validation accuracy of approximately 79%. Figures 10.2 and 10.3 depict the plot of accuracy vs. number of epochs and loss vs. number of epochs for training and validation of the CNN model, respectively.

The basic objective of this study is to understand the effectiveness and importance of the face detection stage in emotion detection. The detailed approach and experimental results are discussed in subsections.

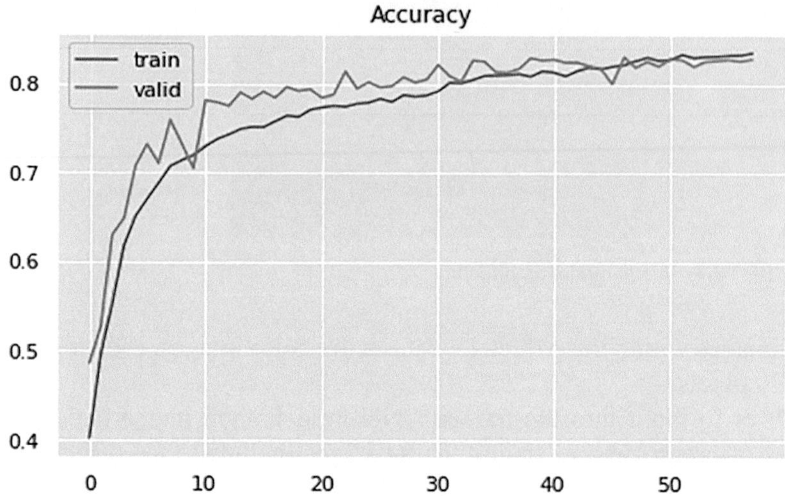

FIGURE 10.2 Plot of accuracy versus number of epochs.

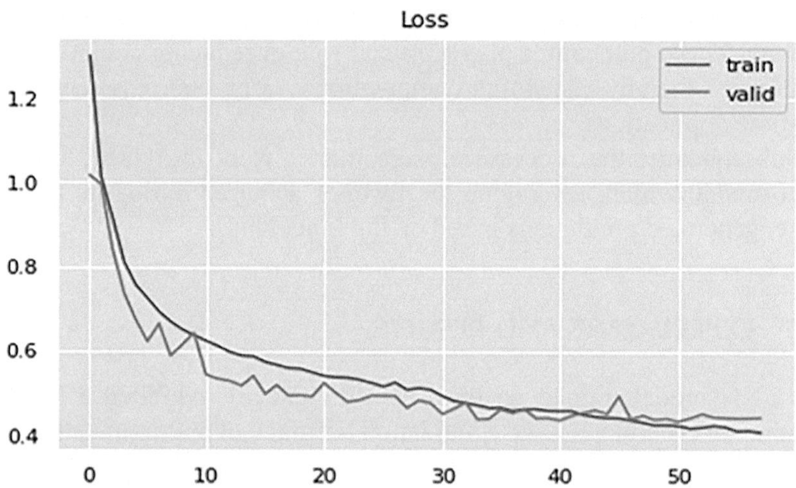

FIGURE 10.3 Plot of loss versus number of epochs.

10.5.1 TESTING

AffectNet is a dataset designed for FER. The dataset includes images with a wide range of ages, ethnicities, and lighting conditions, rendering it suitable for training and testing emotion recognition models in real-world scenarios.

However, for this study, we utilized a subset of the AffectNet dataset, which contains samples collected for different emotions. Each image is a color (3-channel R, G, B) image, with each image measuring 96×96 pixels.

The class-wise distribution of the test samples used from the AffectNet dataset is described below (Table 10.2).

TABLE 10.2 Test Dataset: Number of Samples per Class

Emotion	Number of Samples
Happy	5044
Sad	3091
Neutral	3218

10.5.1.1 Testing without applying face detection

The scaled-down, preprocessed image samples from the AffectNet dataset were fed to the CNN model. The average accuracy of the model on the test dataset was approximately 68%. The confusion matrix for this experiment is presented in Figure 10.4.

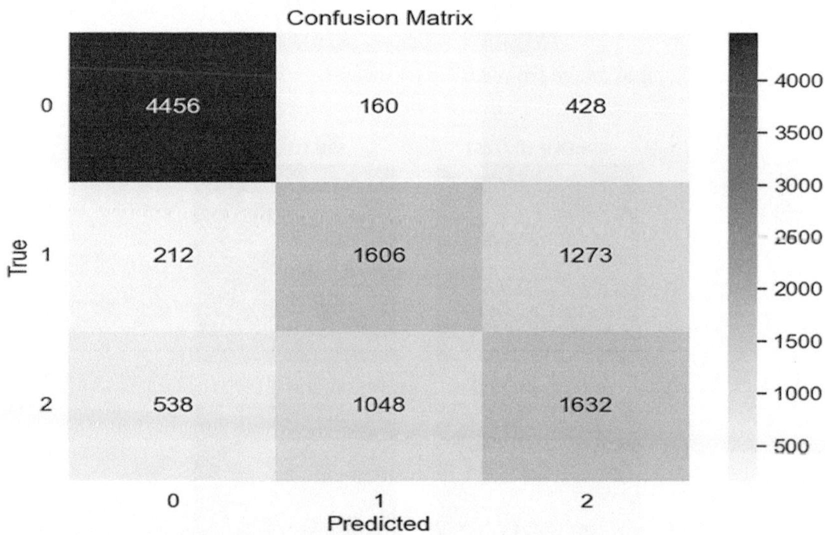

FIGURE 10.4 Confusion matrix for predictions generated by the trained CNN model when tested without applying the face detection algorithm (0: "Happy," 1: "Sad," 2: "Neutral").

10.5.1.2 Testing after applying face detection

Numerous face detection algorithms with diverse underlying architectures are available for consideration. Among the face detection algorithms examined in our experiments are OpenCV (utilizing the Haar Cascade frontal face detector), SSD (single shot multibox detector), MTCNN (multi-task cascaded convolutional network), and RetinaFace.

Our exploration involved a comprehensive examination of the functionality of these four algorithms and an assessment of their suitability for real-time applications, considering factors such as time overhead, face detection accuracy, and precision. Distinct differences in the performance of each algorithm can be observed in the Figures 10.5 and 10.6.

FIGURE 10.5 Sample image from AffectNet dataset.

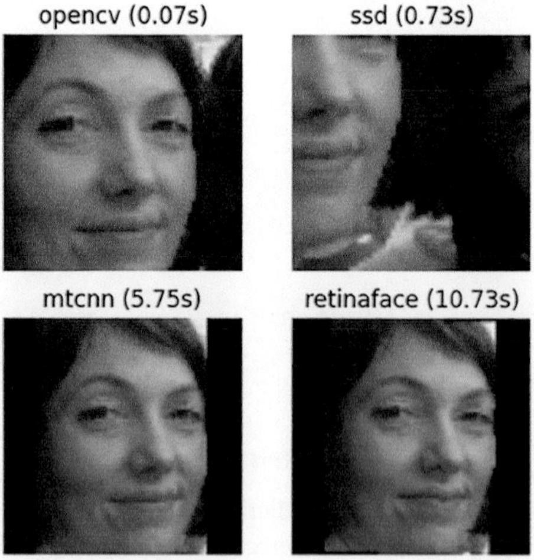

FIGURE 10.6 Results after applying face detection algorithms and cropping the images.

Considering the real-time application of emotion detection and acknowledging the additional time overhead incurred by the use of face detection algorithms and their demand for high computational power, we opted to assess our results using the computationally less expensive OpenCV (Haar Cascade frontal face) detection algorithm. This decision was informed by its ability to capture most, if not all, of the dominant facial sections in an image, proving valuable for feature extraction in the CNN model.

Each input sample underwent a standardized preprocessing procedure involving face detection and subsequent image cropping based on the ROI. The image was then converted to grayscale and resized to dimensions of 48×48 pixels. These preprocessed samples were then fed into the CNN model, and the corresponding confusion matrix depicting the results is provided in Figure 10.7.

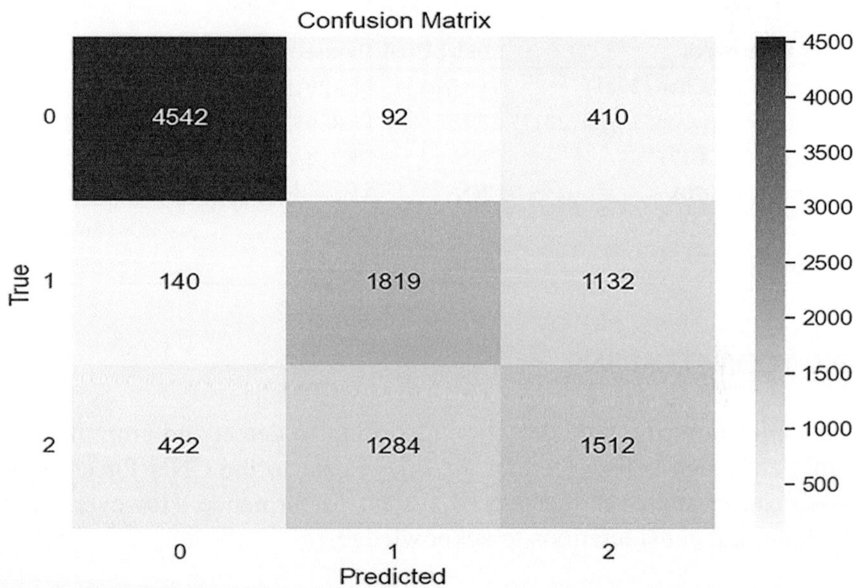

FIGURE 10.7 Confusion matrix for predictions generated by the trained CNN model when tested after applying the OpenCV Haar Cascade face detection algorithm (0: "Happy," 1: "Sad," 2: "Neutral").

The overall results obtained are summarized in Table 10.3.

TABLE 10.3 Results of Experimentation

Emotion	Testing Accuracy (Without Face Detection)	Testing Accuracy (After Applying Face Detection)
Happy	88.3%	90%
Sad	51.9%	58.54%
Neutral	50.71%	47%
Overall Accuracy	68%	69%

Table 10.4 depicts the comparison of the proposed system with previously implemented systems referred to in the study.

It must be noted that our system classifies only three emotions, namely: happy, sad, and neutral, whereas many of the referred systems classify additional emotions such as happiness, sad, neutral, fear, anger, disgust, and so forth.

TABLE 10.4 Comparison with Referred Systems

Paper References	Model Used	Dataset	Accuracy
Khaireddin and Chen (2021)	VGGNet	FER2013	73.28%
Tümen, Söylemez, and Ergen (2017)	CNN	FER2013	57.1%
Pitaloka et al. (2017)	CNN	CK+, JAFFE, MUG	93.14%
Pranav et al. (2020)	CNN	Manually collected images	78.04%
Reyes et al. (2020)	SVM	CK+	86%
Proposed system	CNN	AffectNet	69%

10.6 CONCLUSIONS

The application of a face detection algorithm to detect and crop the face within an image before feeding the image data to the CNN for emotion classification improves the overall system performance. However, there are important considerations to acknowledge.

10.6.1 THE CNN MODEL PERFORMANCE

10.6.1.1 Quality of training data

The CNN model utilized in the backend was exclusively trained on the top three most populated classes to avoid overfitting while ensuring the model learned relevant features from a sizable dataset. The model's ability to extract the most useful and relevant features heavily depends on the quality of input data used during the training phase. Misclassified data and uneven distribution of available data samples can cause the model to learn incorrect or misleading features.

10.6.1.2 Model training

Employing data augmentation, optimizers, and callbacks during training prevented overfitting and guided the model away from local minima in the loss function. This approach is validated by the high training and validation accuracies (*see* Figures 10.2 and 10.3). The model's generalization capabilities are further demonstrated when tested on the unseen AffectNet dataset (*see* Figures 10.4 and 10.7).

10.6.2 IMPROVEMENT IN PERFORMANCE

The model performs better than the estimated average human accuracy on the FER2013 dataset, which is around 65.5% (Goodfellow et al., 2013). The face detection stage before testing helps eliminate irrelevant background information and noise, allowing the underlying CNN model to focus on important features and generate comparatively better results on the unseen dataset.

10.6.3 LIMITATIONS OF THE HAAR CASCADE ALGORITHM

The Haar Cascade face detection algorithm, part of OpenCV, has limitations—particularly in its suitability for frontal face images. In real-world scenarios, variations in face position, orientation, and alignment can hinder its effectiveness. This algorithm relies heavily on pixel intensity differences, potentially resulting in false face detections.

Comparatively, MTCNN and RetinaFace offer better face detection accuracy, but the Haar Cascade algorithm significantly reduces time overhead, which is crucial for real-time applications.

In conclusion, applying face detection and extracting relevant features improves model performance. Although face detection contributes to emotion recognition, the methods implemented in this study are optimized for efficient classification of frontal face images only. Thus, there is scope for improvement through methodologies that can handle images with varying face orientations.

To better address real-world scenarios, additional factors such as varying light intensities, facial positions, and so on, should be considered. The refinement of model performance hinges on its ability to extract the most relevant features—a goal further supported by robust face detection techniques and more advanced feature extraction methods. This collaboration strengthens model performance and lays the foundation for a robust and precise emotion recognition system, capable of excelling in a wide array of complex real-world applications.

DECLARATION

No AI tools or any other similar tools have been used in the content creation of this chapter.

KEYWORDS

- **convolutional neural network**
- **data augmentation**
- **emotion detection**
- **face detection**
- **Haar-Cascade algorithm**
- **machine learning algorithm**
- **OpenCV**
- **pixel intensity**
- **real-time systems**
- **retina face**

REFERENCES

Ekman, P., & Friesen, W. V. (1971). Constants across cultures in the face and emotion. *Journal of Personality and Social Psychology, 17*(2), 124.

Goodfellow, I. J., Erhan, D., Carrier, P. L., Courville, A., Mirza, M., Hamner, B., & Bengio, Y. (2013). Challenges in representation learning: A report on three machine learning contests. In *Neural Information Processing: 20th International Conference, ICONIP 2013, Daegu, Korea, November 3–7, 2013. Proceedings, Part III, 20* (pp. 117–124). Springer Berlin Heidelberg.

Gowri, S. M., Rafeeq, A., & Devipriya, S. (2021). Detection of real-time facial emotions via deep convolution neural network. In *2021 5th International Conference on Intelligent Computing and Control Systems (ICICCS)* (pp. 1033–1037). IEEE.

Haghpanah, M. A., Saeedizade, E., Masouleh, M. T., & Kalhor, A. (2022). Real-time facial expression recognition using facial landmarks and neural networks. In *2022 International Conference on Machine Vision and Image Processing (MVIP)* (pp. 1–7). IEEE.

Jain, D. K., Shamsolmoali, P., & Sehdev, P. (2019). Extended deep neural network for facial emotion recognition. *Pattern Recognition Letters, 120*, 69–74.

Khaireddin, Y., & Chen, Z. (2021). Facial emotion recognition: State of the art performance on FER2013. *arXiv preprint*, arXiv:2105.03588. https://arxiv.org/abs/2105.03588 (accessed on 11th July 2025).

Krizhevsky, A., Sutskever, I., & Hinton, G. E. (2017). ImageNet classification with deep convolutional neural networks. *Communications of the ACM, 60*(6), 84–90.

Mehendale, N. (2020). Facial emotion recognition using convolutional neural networks (FERC). *SN Applied Sciences, 2*(3), 446.

Pitaloka, D. A., Wulandari, A., Basaruddin, T., & Liliana, D. Y. (2017). Enhancing CNN with preprocessing stage in automatic emotion recognition. *Procedia Computer Science, 116*, 523–529.

Pranav, E., Kamal, S., Chandran, C. S., & Supriya, M. H. (2020). Facial emotion recognition using deep convolutional neural network. In *2020 6th International Conference on Advanced Computing and Communication Systems (ICACCS)* (pp. 317–320). IEEE.

Reyes, S. R., Depano, K. M., Velasco, A. M. A., Kwong, J. C. T., & Oppus, C. M. (2020). Face detection and recognition of the seven emotions via facial expression: Integration of machine learning algorithm into the Nao robot. In *2020 5th International Conference on Control and Robotics Engineering (ICCRE)* (pp. 25–29). IEEE.

Tang, Y. (2013). Deep learning using linear support vector machines. *arXiv preprint*, arXiv:1306.0239. https://arxiv.org/abs/1306.0239 (accessed on 11th July 2025).

Tümen, V., Söylemez, Ö. F., & Ergen, B. (2017). Facial emotion recognition on a dataset using convolutional neural network. In *2017 International Artificial Intelligence and Data Processing Symposium (IDAP)* (pp. 1–5). IEEE.

CHAPTER 11

Deep Learning-Based Song Recommendation System Using Facial Expression

SHEELA CHINCHMALATPURE,* TANUSHREE PRAKASH KANADE, PURVA SARVADE, GAURAV ZANWAR, SHARVARI DHAGE, ASHITOSH WANKHADE, and TANUJ SOMANI

Department of Artificial Intelligence and Data Science, BRACT's Vishwakarma Institute of Technology, Pune, Maharashtra, India

*Corresponding author

ABSTRACT

In today's music streaming landscape, personalized song recommendations are crucial for enhancing user experience. This research presents a novel approach using deep learning and facial expression recognition to recommend songs based on users' emotional states. A convolutional neural network (CNN) model with 92% accuracy is employed to analyze facial expressions and map them to emotions like happiness, sadness, excitement, or calmness. These identified emotional states are then used to select songs from a music library tagged with emotional metadata. By aligning music recommendations with real-time emotional cues, the system ensures a more meaningful and emotionally resonant listening experience. This innovative integration of visual emotion detection and music personalization enhances user engagement and sets a new direction for emotionally intelligent music discovery.

Intelligent Computing and Analytics in Image Processing. Samarjeet Borah, Jambi Ratna Raja Kumar, Laxmi Bewoor, Rahul Raghvendra Joshi, and Mahendra Prabhakar Deore (Eds.)

DOI: 10.1201/9781779640772-11

11.1 INTRODUCTION

In the rapidly changing environment of modern music streaming, the delivery of personalized song recommendations has become an indispensable cornerstone in creating a captivating and highly individualized user experience. In a world where an extensive catalog of music is available at our fingertips, the challenge lies not in offering an overwhelming number of options but in intelligently guiding users toward the music that resonates most profoundly with their unique preferences and emotional states. Traditional recommendation systems predominantly rely on historical listening data and song metadata, often falling short in capturing the dynamic and nuanced nature of human emotions intricately linked to the appreciation of music.

This research work ventures into uncharted territory by introducing a groundbreaking and innovative approach to song recommendations, integrating deep learning (DL) and real-time facial expression recognition. Music is a profound vehicle for emotional expression and connection, and it is imperative to comprehend a user's real-time emotional state to deliver recommendations that are genuinely personalized.

In our implementation, we steer away from pre-trained models and develop a custom convolutional neural network (CNN). This CNN model exhibits an impressive accuracy rate of 92% in real-time emotion detection through the analysis of facial expressions. By doing so, we eliminate the one-size-fits-all approach and ensure that the model is finely tuned to capture the subtle nuances of emotions, regardless of individual differences.

Once emotions are accurately recognized in real-time, we employ an API-based approach to recommend songs that align seamlessly with the user's dynamically detected emotional state. Our system interfaces with a vast music library, ensuring that the suggested songs are not only contextually relevant but also emotionally harmonized with the user's immediate state of mind.

This innovative approach promises to significantly enhance the music listening experience, enabling users to discover and enjoy music that not only matches their preferences but also resonates with their ever-evolving emotional state. This research paper delves into the theoretical foundations of the study, the intricate technical details of our custom-built CNN model, the real-time facial expression recognition methodology, and the workings of the API-based song recommendation system.

Furthermore, this research explores the broader implications for music streaming platforms and user engagement. By uniquely integrating artificial intelligence (AI), music, and emotional intelligence, our research work seeks to foster a deeper and more personal connection between users and their music. This endeavor, accomplished without pre-trained models, exemplifies the potential of custom-built solutions to transform the landscape of music streaming recommendations, enhancing the overall listening experience in profound ways.

11.2 METHODOLOGY

A facial expression-driven song recommendation system using a CNN for real-time emotion-based suggestions (Figure 11.1).

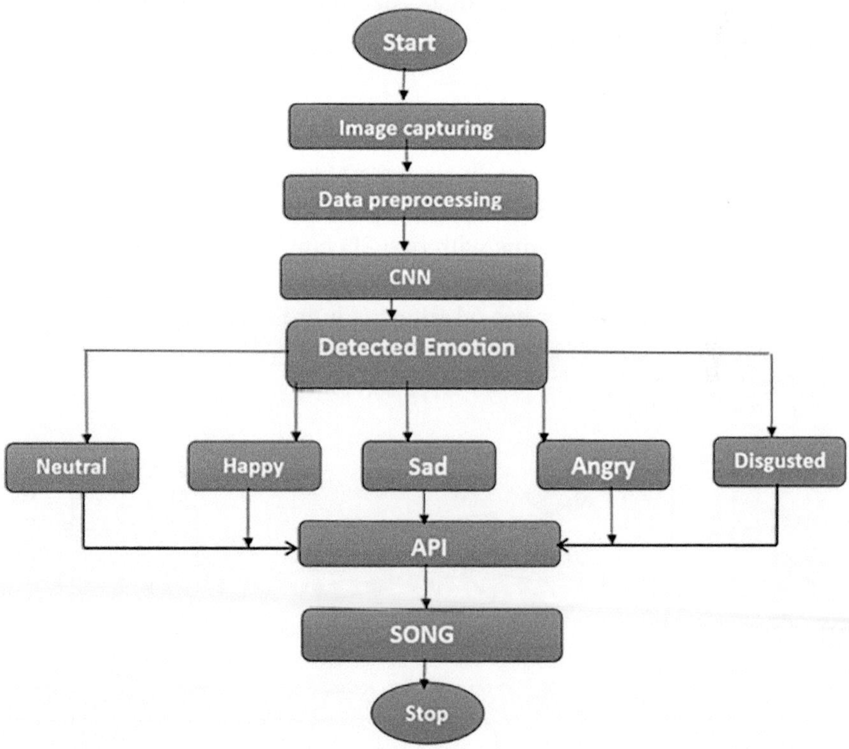

FIGURE 11.1 System flow process.

11.2.1 DATA COLLECTION

We collect facial expression data in real time using an integrated webcam and train it with the FER2013 dataset to enhance our system's ability to suggest tailor-made songs according to the current emotions of users, thereby creating a more personalized music experience.

11.2.2 PREPROCESSING STEPS

We used the flow_from_directory technique for structured batch processing and ImageDataGenerator for effective data augmentation in our research to fully utilize the capabilities of the Keras library. Each test and training image was standardized as part of the data preprocessing pipeline, which included grayscale conversion and scaling to a consistent 48×48 pixel size. This careful preparation improves the overall research findings by ensuring consistency in the image data and optimizing it for further analysis and model training.

11.2.3 FEATURE EXTRACTION USING CNN

In the framework of emotion analysis, a CNN model is designed to extract features from facial images. The conceptual design consists of several essential components. It begins with two 2D convolutional layers that use rectified linear unit (ReLU) activation functions. The initial layer has 32 filters, and the second layer has 64 filters (Figure 11.2).

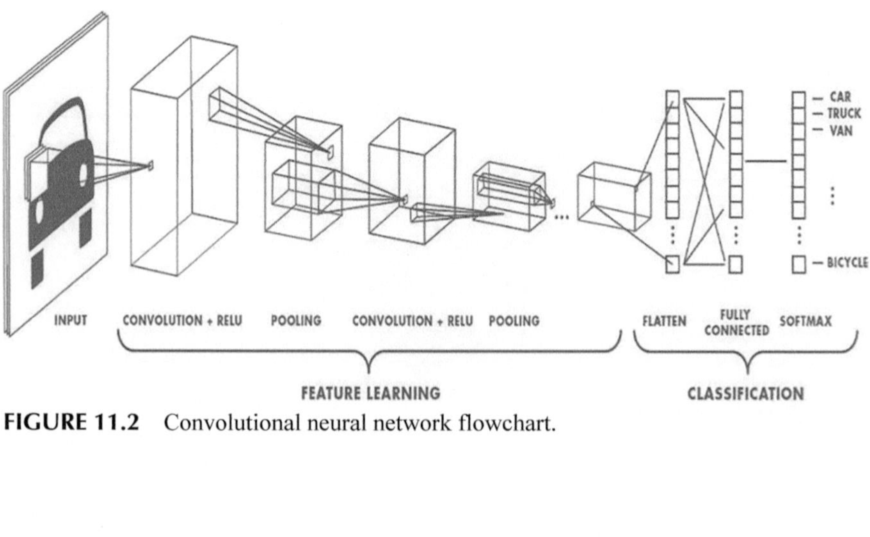

FIGURE 11.2 Convolutional neural network flowchart.

➢ **Two-Dimensional Convolution Operation Formula:**

$$O[i,j] = \sum_{n}\sum_{m} I[i+n, j+m]\cdot K[m,n]$$

➢ **Rectified Linear Unit (ReLU) Activation Function:**

$$f(x) = \max(0,x)$$

The goal of these convolutional layers is to capture complex patterns found in facial expressions. To minimize spatial dimensions, max-pooling layers with 2×2 pool sizes are added. To avoid overfitting, dropout layers with a dropout rate of 0.25 are introduced.

Two additional 2D convolutional layers, each with 128 filters, along with max-pooling layers for dimensionality reduction, help the network better understand facial expressions.

➢ **Max Pooling Operation Formula:**

$$O[i,j] = \max(I[i\cdot\text{stride}: i\cdot\text{stride} + \text{pool\_size}, j\cdot\text{stride}: j\cdot\text{stride} + \text{pools\_size}])$$

Model resilience is further improved by adding a final dropout layer with a 0.5 dropout rate. Using a softmax activation function, the output layer's seven units represent the seven distinct emotions that need to be classified.

➢ **Softmax Function for Probability Calculation:**

$$P(y = i \mid X) = (e^{x}{}_{i}) \div (\textstyle\sum_{j} e^{x}{}_{j})$$

The CNN model is built using a learning rate of 0.0001. The Adam optimizer and the categorical cross-entropy loss function are utilized. The model is trained using training and validation data generators. After training, the learned weights are saved in an .h5 file for further usage, and the model structure is saved in a .json file.

This CNN model forms the heart of the emotion analysis system. It efficiently extracts features from facial images and classifies emotions using those features.

11.3 EXPERIMENTAL OUTPUT

In our system's concluding phase, we leverage the CNN model's capabilities to accurately evaluate users' emotional states through their facial expressions. This emotion, diligently discerned by our CNN, constitutes a pivotal component within our song recommendation framework.

We seamlessly integrate an application programming interface (API) to establish a connection between the recognized emotion and an extensive song library. This API serves as the conduit between the real-time detected emotion and a thoughtfully curated selection of songs that harmonize with that specific emotional context.

This dynamic procedure enables us to correlate the recognized emotions with a diverse collection of songs, ensuring that our song recommendations align seamlessly with the user's prevailing emotional disposition. The outcome is a personalized playlist that resonates with the user's feelings, elevating their music listening experience by offering songs that authentically mirror their emotional state.

11.4 RESULTS AND DISCUSSION

11.4.1 CNN MODEL'S OUTPUT

We used the FER2013 dataset, which is composed of facial images categorized into seven different emotion classes: neutral, surprise, happiness, sadness, disgust, and anger. For model training, hyperparameter tuning, and evaluation, the dataset was divided into subsets for training (80%), validation (10%), and testing (10%).

The emotion detection model demonstrated its ability to identify facial emotions with an accuracy of 92% on the test set. Sample outputs can be seen in Figure 11.3.

FIGURE 11.3 Output images.

11.4.2 FINAL OUTPUT

Based on the identified facial expressions, the recommendation system—connected with the Spotify API—provided music recommendations. Screenshots of the results are presented in Figure 11.4.

```
Playlist: Good Mood Happy Mix - spotify:playlist:37i9dQZF1EIgNoWOvbnUCk
Playlist: Mood Booster - spotify:playlist:37i9dQZF1DX3rxVfibe1L0
Playlist: Feel Good Piano - spotify:playlist:37i9dQZF1DXcEKFjZJYZcc
Playlist: Happy Hits! - spotify:playlist:37i9dQZF1DXdPec7aLTmlC
Playlist: Wake Up Happy - spotify:playlist:37i9dQZF1DX0UrRvztWcAU
Playlist: Happy Pop Hits - spotify:playlist:37i9dQZF1DWVlYsZJXqdym
Playlist: Mood Booster: the Happy Playlist - spotify:playlist:0IAG5sPikOCo5nvyKJjCYo
Playlist: Happy Favorites - spotify:playlist:37i9dQZF1DWZKuerrwoAGz
Playlist: Happy Songs ☺ (80s, 90s, 2000s, 2010s & 2020s) ☺ Best Happy Mood  - spotify:playlist:4AnAUkQNrLKlJCInZGSXRO
Playlist: Happy Pop - spotify:playlist:37i9dQZF1DX1H4LbvY4OJi
Track: Mood Happy - spotify:track:0tK7IB1wh5Oeuu8lYm9Pth
Track: Good Mood - Original Song From Paw Patrol: The Movie - spotify:track:4uX1pkSuSidzJxT4eWL7x1
Track: This Love - spotify:track:5nGn71GSdopFmAiZ2dmG38
Track: Good Mood - spotify:track:2PB428pRLBrW2IONANRbB1
Track: Sunday Morning - spotify:track:3588Rm1WkQMFqH2YmfjRS4
Track: For All We Know - spotify:track:086RZ0U22i29HR5bj0sWSB
Track: Dog Days Are Over - spotify:track:3QWxAUn4kc8Xhkqa4C2dcg
Track: Mood - spotify:track:3g6C863vEPs7H14SUGJAhj
Track: Better Together - spotify:track:581fmmD1CXx8oGoSBPyZ1v
Track: Good Mood - spotify:track:3Kr8aPFBZ0Vs3ZcOyGBcGM
PS C:\Users\Tanu\Desktop\Emotion_detection_with_CNN\Emotion_detection_with_CNN-main>
```

Happy

```
Playlist: Moody Sad Mix - spotify:playlist:37i9dQZF1EIhmSBwUDxg84
Playlist: Jazz Noir - spotify:playlist:37i9dQZF1DWWR73B3Bnjfh
Playlist: Sad Classical - spotify:playlist:37i9dQZF1DXbm0dp7JzNeL
Playlist: kukon sad mood - spotify:playlist:24eFMtBo4dfLRcFzMy4UEs
Playlist: Sad tiktok mood - spotify:playlist:71K3ucoKFsGSjsskYL4xNd
Playlist: Dark Academia Jazz - spotify:playlist:37i9dQZF1DX0J6DFF0p3Vs
Playlist: Melancholy Instrumentals - spotify:playlist:37i9dQZF1DWZrc3lwvImLj
Playlist: mood : sad - spotify:playlist:3Kz5KBE3Ksupz9odBGwze6
Playlist: a sad chet baker mood - spotify:playlist:6G36GWNj57ajAol31t2QzB
Playlist: Mood:SAD - spotify:playlist:0FD0lmtg8rzHU5sKb1Ocfk
Track: Modo Sad - spotify:track:7k6kbhO4J3wCm9PLD6kvpx
Track: Mood - Radio Edit - spotify:track:7ADPf4TMrbMR1ZieZnO5TB
Track: Mood - Radio Edit - spotify:track:1ulTVBUHDTJC9OWJsh9bs4
Track: Mood Bad - spotify:track:0VCPSdri6GObtK0Wln5vcb
Track: Mood (Say My Name) - spotify:track:25iqW8Z4FdA14gOsnsByzO
Track: Moon - Sad Lofi Rain version - spotify:track:0gluM2FjwJFxpTfwfPLFOp
Track: A Voice in My Head and Rainy - spotify:track:4odjnu3I8aQyYmw5dTpEw4
Track: SAD! - spotify:track:7J2NL5JOFdVG3qElbj0KjQ
Track: Dance Monkey - spotify:track:5K8cRD6pNj7nvTkuf7ZOJ8
Track: Sad Mood - spotify:track:1XYNrxsbgkDdPtTQLJhqNd
PS C:\Users\Tanu\Desktop\Emotion_detection_with_CNN\Emotion_detection_with_CNN-main>
```

Sad

```
Playlist: Angry Moody Mix - spotify:playlist:37i9dQZF1EIg1pTZ3BUIEF
Playlist: Angry mood song tamil 😡 - spotify:playlist:3p8ejB7BscAVmEdyK7AtXx
Playlist: Mood:Angry - spotify:playlist:41ZY4409K5lr3aJYq4JX95
Playlist: angry mood hindi songs - spotify:playlist:3JNWpteYvH3ynMcyPcvxfx
Playlist: Angry Mood 😡 - spotify:playlist:15OOgBSCBRPbkcYm9drzjE
Playlist: Mood:Angry  - spotify:playlist:2fpK20IB1TfZgMugu2xtM6
Playlist: Angry mood😡 - spotify:playlist:2OntVb3hPdWcxWRreKW9z6
Playlist: Angry Mood 😡 - spotify:playlist:2yWAdmIrf0rbchGJlJNwsx
Playlist: Angry mood vibes - spotify:playlist:6LnvGbJH3M8Zy831jmheBn
Playlist: Angry mood 😡 - spotify:playlist:5ntJ2agws0K7YprQqOpm2p
Track: Angry Mood - spotify:track:2n4Ev6ip5b1IcLC5dAQndZ
Track: Angry Mood - spotify:track:4rb2hnzWIzMsPLlz0Jaadl
Track: Angry Mood - spotify:track:086JFYFADOa7crJTNkZx2J
Track: Angry Mood - spotify:track:5m3Nojvx51Ehw5LCd2692C
Track: Angry Mood - spotify:track:5ZvOoc0cDf1Gg55xA4OqH3
Track: Angry Mood - spotify:track:1Pc5uO8Y3OPFWa4Sv9hbk9
Track: Angry Mood - spotify:track:3GfgeHFxNZVfWTqKG5BGiz
Track: Angry Mood - spotify:track:4WPtYPugivYeib0DKs38oa
Track: Angry Mood - spotify:track:5q4pFkRptLmy0b0CAHoebb
Track: Angry Mood - spotify:track:2ipRun6o9Pjc9MTuPaeRKn
PS C:\Users\Tanu\Desktop\Emotion_detection_with_CNN\Emotion_detection_with_CNN-main>
```

Angry

(Continued)

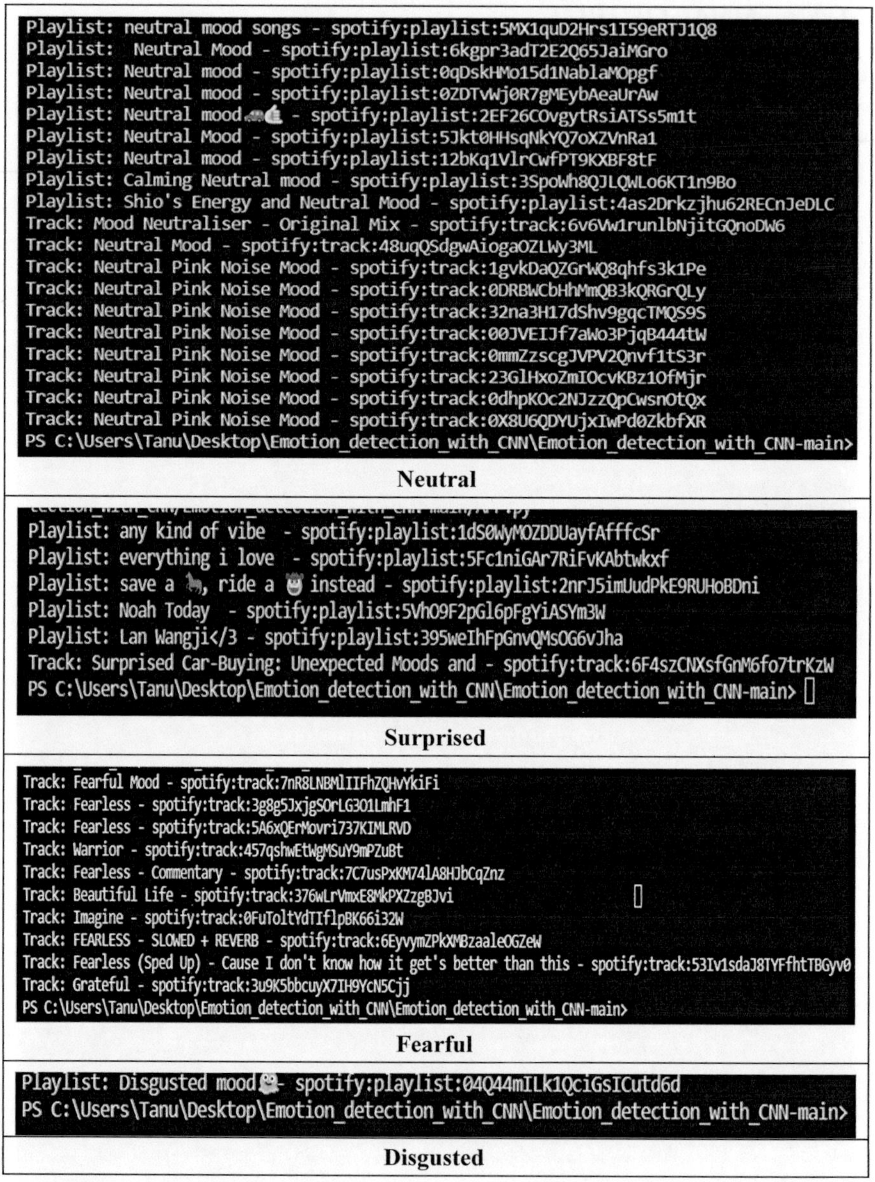

FIGURE 11.4 Results snapshots.

According to user feedback surveys, 90% of users were satisfied, confirming that the recommended songs matched the emotional context of users, as interpreted from their facial expressions (Table 11.1 and Figure 11.5).

$$\text{Accuracy} = (TP + TN)/(TP + TN + FP + FN)$$

where; TP is the true positive; TN is the true negative; FP is the false positive; and FN is the false negative.

TABLE 11.1 Comparison of Different Algorithms

Algorithm	CNN	DT	MF	RF	SVM
Accuracy	85%	70%	75%	80%	78%
Computational resources	Higher	Lower	Modest	Modest	Modest
Data requirements	Larger	Lower	Modest	Lower	Modest
Versatility	Higher	Lower	Lower	Modest	Modest

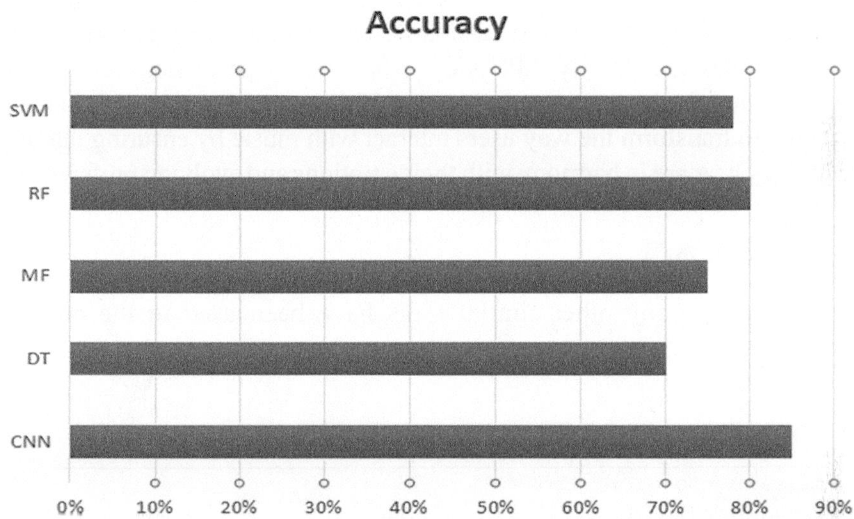

FIGURE 11.5 Accuracy of different algorithms.

11.5 CONCLUSION

In brief, this study integrates facial expression analysis using CNNs and the Spotify API to offer a novel approach to music recommendation systems. It demonstrates how using facial expressions as non-verbal input can enhance personalization and increase user satisfaction and engagement.

By utilizing CNN, the system accurately recognizes emotional states by effectively extracting significant features from facial expressions. The inclusion of the Spotify API allows the system to generate music recommendations that align with users' real-time emotional states. This adaptive and dynamic recommendation strategy represents a major step forward in personalized content delivery.

Our experiments show that, in terms of user satisfaction and the relevance of song suggestions, the facial expression-based music recommendation system outperforms traditional approaches. Users reported feeling more emotionally connected and engaged with the music, and recommendations consistently reflected their changing emotional states in real time.

However, it is important to acknowledge certain limitations, such as privacy concerns related to facial expression data and the need for broader user testing. Future work may focus on enhancing data security, improving the model's ability to recognize subtle emotions, and incorporating additional contextual factors to further refine recommendation accuracy.

In conclusion, the facial expression-based song recommendation system that integrates the Spotify API and CNN offers a promising solution to improve music consumption experiences. This innovative approach has the potential to transform the way users interact with music by ensuring that the songs they hear are in harmony with their emotions and evolving preferences.

DECLARATION

No AI tools or any other similar tools have been used in the content creation of this chapter.

KEYWORDS

- **application programming interface**
- **convolutional neural network**
- **deep learning**
- **emotional state**
- **facial expression analysis**
- **music discovery**
- **real-time emotion detection**
- **rectified linear unit**
- **user engagement**
- **user satisfaction**

REFERENCES

Jaison C, R. R., & Rajeswari, M. (2023). Song recommendation based on voice tone analysis. *2023 Second International Conference on Electronics and Renewable Systems (ICEARS)*, Tuticorin, India, pp. 708–712.

Kodama, Y. (2005). A music recommendation system. *2005 Digest of Technical Papers. International Conference on Consumer Electronics, 2005. ICCE.*, Las Vegas, NV, pp. 219–220. https://doi.org/10.1109/ICCE.2005.1429796 (accessed on 11th July 2025).

Kumar, R., & Rakesh. (2022). Music recommendation system using machine learning. *2022 4th International Conference on Advances in Computing, Communication Control and Networking (ICAC3N)*, Greater Noida, India, pp. 572–576. https://doi.org/10.1109/ICAC3N56670.2022.10074362 (accessed on 11th July 2025).

Lin, Y. (2022). Design and implementation of music recommendation system based on big data platform. *2022 IEEE 2nd International Conference on Computer Systems (ICCS)*, Qingdao, China, pp. 51–54. https://doi.org/10.1109/ICCS56273.2022.9988197 (accessed on 11th July 2025).

Niyazov, E., Mikhailova, E., & Egorova, O. (2021). Content-based music recommendation system. *2021 29th Conference of Open Innovations Association (FRUCT)*, Tampere, Finland, pp. 274–279. https://doi.org/10.23919/FRUCT52173.2021.9435533 (accessed on 11th July 2025).

Patel, & Wadhvani, R. (2018). A comparative study of music recommendation systems. *2018 IEEE International Students' Conference on Electrical, Electronics and Computer Science (SCEECS)*, Bhopal, India, pp. 1–4. https://doi.org/10.1109/SCEECS.2018.8546852 (accessed on 11th July 2025).

Sarin, E., Vashishtha, S., Megha, & Kaur, S. (2022). SentiSpotMusic: A music recommendation system based on sentiment analysis. *2021 4th International Conference on Recent Trends in Computer Science and Technology (ICRTCST)*, Jamshedpur, India, pp. 373–378. URL (accessed on 11th July 2025).

Sharma, V. P., Gaded, A. S., Chaudhary, D., Kumar, S., & Sharma, S. (2021). Emotion-based music recommendation system. *2021 9th International Conference on Reliability, Infocom Technologies and Optimization (Trends and Future Directions) (ICRITO)*, Noida, India, pp. 1–5. https://doi.org/10.1109/ICRITO51393.2021.9596276 (accessed on 11th July 2025).

Vayadande, K., Narkhede, P., Nikam, S., Punde, N., Hukare, S., & Thakur, R. (2023). Facial emotion-based song recommendation system. *2023 International Conference on Computational Intelligence and Sustainable Engineering Solutions (CISES)*, Greater Noida, India, pp. 240–248. https://doi.org/10.1109/CISES58720.2023.10183606 (accessed on 11th July 2025).

Wishwanath, C. H. P. D., & Ahangama, S. (2019). A personalized music recommendation system based on user moods. *2019 19th International Conference on Advances in ICT for Emerging Regions (ICTer)*, Colombo, Sri Lanka, pp. 1–1. https://doi.org/10.1109/ICTer48817.2019.9023727 (accessed on 11th July 2025).

CHAPTER 12

Sketch to Photo Synthesis Using Autoencoder–Kalman Filter

JYOTI S. RAGHATWAN[1,*] and SANDHYA ARORA[2]

[1]Department of Computer Engineering, Smt. Kashibai Navale College of Engineering, Savitribai Phule Pune University, Maharashtra, India.

[2]Department of Computer Engineering, MKSSS's Cummins College of Engineering for Women, Savitribai Phule Pune University, Maharashtra, India

*Corresponding author

ABSTRACT

Sketch-to-photo synthesis is an image-to-image translation process with several applications, including assisting in suspect identification. It is a difficult task due to the domain differences between sketches and colored photographs. There are several challenges in handling noisy, broken, or corrupted sketches, notwithstanding the substantial amount of research conducted in this field. This chapter proposes an Autoencoder–Kalman Filter framework to generate face images from a given sketch image. The framework's primary goal is to manage noisy, broken, or corrupted sketches. Based on the current state, the Kalman Filter is used to forecast or estimate the upcoming state. This framework learns how to represent data so that it can be effectively used by the Kalman Filter to produce high-quality images. The proposed framework is able to generate perceptually high-quality photos that are closer to the ground truth images.

Intelligent Computing and Analytics in Image Processing. Samarjeet Borah, Jambi Ratna Raja Kumar, Laxmi Bewoor, Rahul Raghvendra Joshi, and Mahendra Prabhakar Deore (Eds.)

DOI: 10.1201/9781779640772-12

12.1 INTRODUCTION

In recent years, sketch-to-photo synthesis has become an extremely active field of research due to its applicability across multiple domains. It remains a highly challenging task because poorly drawn sketches often exhibit significant geometric and spatial distortion. Another major challenge is the domain difference: sketches are black and white and typically lack details about higher-level features.

Therefore, it is difficult to identify a suspect using a sketch drawn by a police artist based on an eyewitness description. Here, sketch-to-photo synthesis proves useful for generating a suspect's photo from their sketch. Recently, several deep learning (DL) models have been proposed by researchers (Chao et al., 2019; Fang et al., 2020; Gao et al., 2012).

Earlier techniques for sketch-to-photo synthesis relied on paired sketch–photo datasets. Most of these traditional methods were based on linear combinations of similar training photo/sketch patches. Gao et al. (2012) proposed a sparse representation-based algorithm, which consists of a sparse neighbor selection (SNS) technique and a sparse-representation-based enhancement technique to synthesize better quality photos and sketches.

The embedded hidden Markov model (EHMM) by Xiao et al. (2009) does not require a large number of training samples. With fewer samples, it generates pseudo-photos. Bayesian Tensor Inference (Güçlütürk et al., 2016) has been used to transfer style between sketch and photo images. To preserve vital details and enhance output quality, a support vector regression-based approach was proposed by Iranmanesh et al. (2018). A multiscale Markov random fields (MRF) model (Jian et al., 2019) was also used to learn facial features at different scales.

All these methods assume frontal pose and neutral expression in the photo–sketch pair. Hence, even small variations in these conditions adversely affect their performance. To overcome this limitation, many recent DL models based on deep convolutional neural networks (DCNNs) (Kazemi et al., 2018) have been proposed. Most recently, generative adversarial networks (GANs) (Lee et al., 2020) have demonstrated strong performance in image generation tasks by selecting appropriate loss functions and generating more realistic and visually pleasing images.

Although existing sketch-to-photo generation methods yield good performance, some limitations remain—such as missing attributes in the output, blurring effects, incorrect features, and poor handling of incomplete sketches.

12.2 RELATED WORK

Deep learning (DL)-based methods have demonstrated their effectiveness in the field of computer vision. These techniques can learn nonlinear spatial transformations between input and output domains. DCNNs have been successfully applied in many cross-modality applications (Li and Wand, 2016; Li-Liang et al., 2015).

Zhang et al. (2011) proposed a six-layer fully convolutional network for photo-sketch generation. In their architecture, a person's identity is preserved using a novel joint generative–discriminative formulation in the optimization of the objective function. Gatys et al. (2016) proposed the first neural style transfer algorithm to transfer the style of a reference image to any given image by learning feature representations at various layers of a CNN.

Li et al. (2015) proposed a model to transfer the style of an image into arbitrary sketch images. They employed a deep neural network for its discriminative power, combined with generative MRF models for texture synthesis. However, some CNN-based methods still have limitations, such as blurred effects in the generated images, which are addressed by GANs.

A GAN consists of two subnetworks: a generator and a discriminator. The generator creates images and attempts to fool the discriminator, which distinguishes between generated and real images. Most researchers focus on network architecture and loss function selection to enhance the quality of synthetic images.

Contextual GAN (Osahor et al., 2020) used two loss functions—perceptual loss and contextual loss—to complete incomplete sketches. The proposed system learns the joint distribution between sketch and photo during training. However, one limitation is that the identity of the generated image is compromised. Zhao et al. (2019) proposed an attribute-guided GAN to preserve the identity of generated images. In their network architecture, a skip-connection approach is used to reduce the number of layers without compromising network performance. The total loss is calculated using reconstruction loss and adversarial loss.

Osahor et al. (2020) proposed a quality-aware GAN architecture to generate multiple images from a single sketch with specific attributes. The composition-aided generative adversarial network (CA-GAN) (Wang and Tang, 2009) was proposed to focus on delicate facial structures and hard-to-generate components by providing a sketch with pixel-wise face labels as input. To increase image quality, a new loss function—compositional reconstruction loss—was introduced in addition to perceptual loss.

Chao et al. (2019) proposed a GAN architecture for high-fidelity face sketch–photo synthesis. It employs a Patch-GAN with residual blocks as the discriminator and a deep residual U-Net as the generator. An effective loss function is created by combining edge feature loss, pixel loss, and high-level loss to generate visually pleasant images.

Zheng et al. (2019) proposed the encoder-guided generative adversarial network (EGGAN) framework, where skip connections are implemented within a cycle-consistent GAN architecture. The feature encoder directs the training phase to enhance the quality of the generated images. Identity preservation is achieved using feature loss and feature consistency loss.

To create high-quality photo images using multi-adversarial networks, the authors explored the CycleGAN framework (Yu et al., 2021), focusing on hidden layers of the generator and applying Cycle Consistency Loss. Kazemi et al. (2018) suggested a Conditional CycleGAN in which the condition is based on facial features, such as skin tone. This technique does not require paired data like sketch–photo pairs for training. Although some properties may be altered in the synthesized photo, the framework still preserves the facial style.

The Identity-Aware CycleGAN (Zhang et al., 2021) was proposed to improve identity preservation in generated faces by emphasizing critical facial features such as the nose and eyes. Recognition accuracy is enhanced using a perceptual loss function.

This chapter suggests a DL model—an Autoencoder with a Kalman Filter—to handle noisy and incomplete sketches. The proposed system is compared with a standard autoencoder model without the Kalman Filter. The suggested approach improves performance in terms of mean square error (MSE) and mean absolute error (MAE). Compared to other state-of-the-art methods, the proposed model generates high-quality photo images that are very close to ground truth images in terms of structural similarity index (SSIM).

12.3 PROPOSED SYSTEM

An autoencoder is a type of neural network trained to learn a compressed representation of input data, such as an image or a signal. It comprises two components: an encoder that maps the input to a lower-dimensional latent space and a decoder that reconstructs the input from this latent space.

The Kalman Filter, proposed by Rudolf E. Kalman, is a mathematical method designed to estimate values that are typically closer to the true values of the measurements. It has been widely used in applications such as image restoration, noise removal, and in space and military systems.

Mathematically, it can be represented as:

Consider $p(x, y)$ as a pixel of an image.

Its surrounding pixels are given by $p(x-m, y-n)$, for $(m, n) \in N$ (range of pixels), and it is represented by:

$$p(x, y) = \Sigma_{(m,n)\in N} \Sigma a_{p,q} \cdot p(x-m, y-n) + u(x, y) \tag{1}$$

where; (m, n) is the coordinate centered on the current pixel; $u(x, y)$ denotes the derived noise. This approach improves noise reduction in the reconstruction of an image.

The detailed architecture of the proposed Autoencoder with Kalman Filter is shown in Figure 12.1. An autoencoder with a Kalman Filter is a framework that combines these two methods to achieve better denoising and state estimation performance. The goal is to learn a mapping between noisy measurements and inputs for the Kalman Filter, which can subsequently determine the system's actual state. While the Kalman Filter can leverage the system's dynamics and past knowledge, the autoencoder can learn the data's relevant characteristics and patterns.

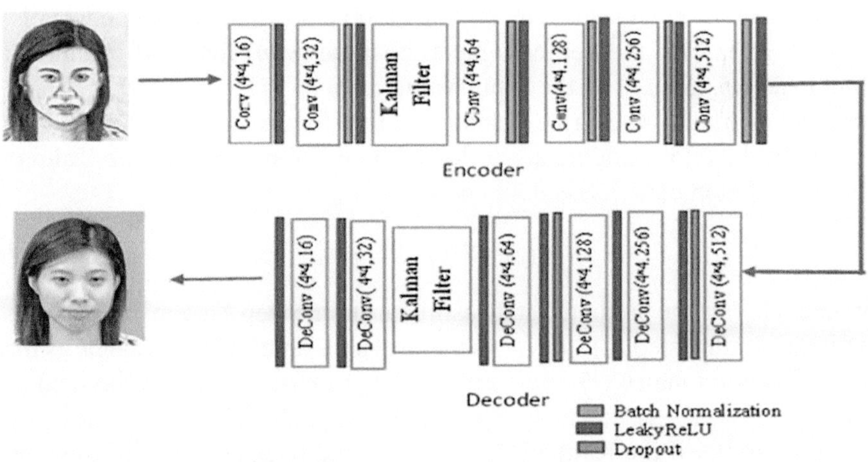

FIGURE 12.1 Autoencoder–Kalman Filter architecture.

12.3.1 KALMAN ENCODER

Let's assume we have an input image of size:

$$H_{in} \times W_{in} \times C_{in}$$

where; H_{in} is the input height; W_{in} is the input width; and C_{in} is the number of input channels.

The desired output feature map size after upsampling is:

$$H_{out} \times W_{out} \times C_{out}$$

where; H_{out} is the output height; W_{out} is the output width; and C_{out} is the desired number of output channels.

The transpose convolution operation can be defined as follows:

- Define the transpose convolution kernel or filter of size $K \times K \times C_{out} \times C_{in}$, where; K is the kernel size.
- Pad the input feature map with appropriate padding to maintain the output size.
- Compute the output feature map using the following equation for each pixel (i, j) in the output feature map:

$$O(i, j, :) = \sum_{m=0}^{K-1} \sum_{n=0}^{K-1} \sum_{c=0}^{C_{in}-1} \text{Kernel}(m, n, :, c) \cdot \text{Input}(i \cdot S + m, j \cdot S + m, j \cdot S + n, c) \quad (2)$$

where; $O(i,j,:)$ represents the output at pixel (i,j) across all channels; Kernel $(m,n,:,c)$ is the kernel value at position (m,n) for input channel c; Input $(i.S+m,j.S+n,c)$ is the input value at pixel $(i.S+m,j.S+n)$ for input channel c; and S is the stride (usually set to 1 for transpose convolution).

Here, the two encoders are followed by a Kalman Filter. The Kalman state prediction equation is defined as:

$$\tilde{\bar{\text{o}}}_{k|k-1} = F \cdot O_{k-1|k-1} + B \cdot u_k \quad (3)$$

where; $\tilde{\bar{\text{o}}}_{k|k-1}$ is the predicted state estimate at time step k given the previous estimate at time step k-1; $O_{k-1|k-1}$ is the previous state estimate; k-1; F is the state transition matrix; B is the control input matrix; and u_k is the control input.

The output of the Kalman Filter is then fed back as input to Equation (2) four times using different filter sizes, as shown in the architecture.

12.3.2 KALMAN DECODE

The output of the Kalman encoder is of size:

$$EH_{in} \times EW_{in} \times EC_{in}$$

where;EH_{in} is the height; EW_{in} is the input width; and EC_{in} is the number of input channels.

The desired output feature map size after downsampling is:

$$DH_{\text{out}} \times DW_{\text{out}} \times DC_{\text{out}}$$

where; DH_{out} is the output height; DW_{out} is the output width; and DC_{out} is the desired number of output channels.

The max pooling operation can be defined as follows:

For each output pixel (i,j) in the output feature map:

$$O(i,j,c) = \max_{m=0}^{K-1} \max_{n=0}^{K-1} \text{Input}(i \cdot S + m, j \cdot S + n, c) \tag{4}$$

where; $O(i,j,c)$ represents the output at pixel (i,j) and channel c; Input($i.S+m,j.S+n,c$) is the input value at pixel $(i.S+m,j.S+n)$ and channel c; K is the pooling kernel size; S is the stride size.

The max pooling operation selects the maximum value within each pooling kernel window, reducing the spatial resolution of the input feature map.

$$\bar{Y}_k = L_{k|k-1} \cdot MM^T \cdot (MM \cdot L_{k|k-1} \cdot MM^T + C)^{-1} \tag{5}$$

where; $Y_{k|k-1}$ is the predicted state estimate at time step k given the previous estimate at time step k-1; MM is the measurement matrix; C is the measurement color covariance matrix; and L is the updated error covariance matrix at iteration k.

The output of the Inverse Kalman Filter is then fed as input to Equation (4) twice, using different filter sizes, as shown in the architecture

12.4 EXPERIMENTAL RESULT

This section provides a summary of the dataset utilized, the training information, and the qualitative and quantitative outcomes.

12.4.1 DATASET AND TRAINING DETAILS

The proposed method was evaluated on CUFS, the CUHK Face Sketch Database (Zheng et al., 2019). There are 188 photo–sketch pairs in the CUHK student dataset. A testing set of 100 photo–sketch pairs and a training set of 88 photo–sketch pairs are included. Due to the small size of the training set, several transformation techniques were applied to enhance the dataset. The training set was expanded to 1800 photo–sketch pairs by including faces that were cropped, shifted, and rotated at various angles. During model training, a 256×256 input image was provided for each test case. The model was trained using the Adam optimizer (Kingma and Ba, 2014), with the learning rate set to 0.001. A total of 200 epochs were used to train the model.

12.4.2 QUANTITATIVE EVALUATION

The suggested approach was evaluated against an autoencoder model without a Kalman Filter. Quantitative indicators like the SSIM, Mean Squared Error (MSE), and MAE were used to compare the generated photo with the ground truth photo and assess the efficacy of the proposed system. Table 12.1 presents a summary of the outcomes of the proposed system. We were able to reduce the MSE and MAE while achieving an average similarity index of up to 82.59%. Table 12.2 presents a summary of average SSIM scores of different GAN-based sketch-to-photo synthesis methods (Figure 12.2).

TABLE 12.1 Comparative Quantitative Measures

Methods	SSIM (%)	MSE	MAE
Autoencoder with Kalman Filter	82.59	3.58	2.419
Autoencoder	69.38	17.39	15.937

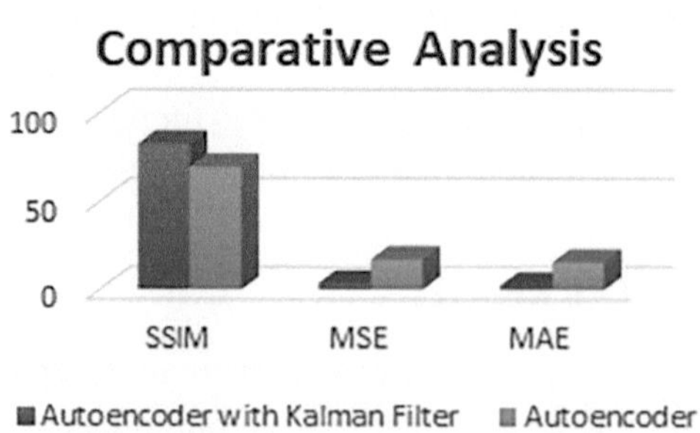

FIGURE 12.2 Comparative analysis.

TABLE 12.2 Summary of Average SSIM Scores of Sketch–Photo Synthesis by Different Methods

Methods	SSIM
Zhao et al. (2019)	0.56
Zheng et al. (2019)	0.65
Wang et al. (2018)	0.79
Fang et al. (2020)	0.64
Proposed Autoencoder–Kalman Filter	0.82

12.4.3 QUALITATIVE EVALUATION

This section presents the qualitative outcomes of the proposed system. Figure 12.3 shows the result of a well-drawn sketch, where we were able to reach a similarity index of up to 84.25%. Even when provided with an incomplete or broken sketch—as in Figure 12.4, where the chin and eyebrow parts are missing—the system was still able to generate a photo with 84.04% SSIM, which reflects only a minimal drop in similarity.

FIGURE 12.3 Result of neatly drawn sketch using Autoencoder–Kalman Filter.

FIGURE 12.4 Result of broken or corrupted sketch using Autoencoder–Kalman Filter.

12.5 CONCLUSION

This chapter presents an Autoencoder–Kalman Filter framework for sketch–photo synthesis. The proposed method explores the ability of the Kalman Filter to handle various forms of smooth curves with different types of noise. Evaluation was performed on the widely used CUFS dataset, and the accuracy of photo–sketch synthesis was significantly improved using the proposed approach in terms of both quantitative and qualitative measures. The proposed method is able to generate visually appealing images, which can be applied in many applications. In future work, the suggested approach may also be applied to other image synthesis problems.

DECLARATION

No AI tools or any other similar tools have been used in the content creation of this chapter.

KEYWORDS

- **autoencoder**
- **deep learning**
- **Kalman filter**
- **mean absolute error**
- **mean squared error**
- **photo-sketch synthesis**
- **quantitative indicators**
- **structural similarity index**

REFERENCES

Chao, W., Chang, L., Wang, X., Cheng, J., Deng, X., & Duan, F. (2019). High-fidelity face sketch-to-photo synthesis using generative adversarial network. https://doi.org/10.1109/icip.2019.8803549 (accessed on 11th July 2025).

Fang, Y., Deng, W., Du, J., & Hu, J. (2020). Identity-aware CycleGAN for face photo-sketch synthesis and recognition. *Pattern Recognition, 102*, 107249. https://doi.org/10.1016/j.patcog.2020.107249 (accessed on 11th July 2025).

Gao, X., Wang, N., Tao, D., & Li, X. (2012). Face sketch–photo synthesis and retrieval using sparse representation. *IEEE Transactions on Circuits and Systems for Video Technology, 22*(8), 1213–1226. https://doi.org/10.1109/tcsvt.2012.2198090 (accessed on 11th July 2025).

Gatys, L. A., Ecker, A. S., & Bethge, M. (2016). Image style transfer using convolutional neural networks. https://doi.org/10.1109/cvpr.2016.265 (accessed on 11th July 2025).

Goodfellow, I., Pouget-Abadie, J., Mirza, M., Xu, B., Warde-Farley, D., Ozair, S., Courville, A., & Bengio, Y. (2017). GAN (generative adversarial nets). *Chinou to Jouhou, 29*(5), 177. https://doi.org/10.3156/jsoft.29.5_177_2 (accessed on 11th July 2025).

Güçlütürk, Y., Güçlü, U., Van Lier, R., & Van Gerven, M. A. J. (2016). Convolutional sketch inversion. In *Lecture Notes in Computer Science* (pp. 810–824). https://doi.org/10.1007/978-3-319-46604-0_56 (accessed on 11th July 2025).

Iranmanesh, S. M., Dabouei, A., Kazemi, H., & Nasrabadi, N. M. (2018). Deep cross polarimetric thermal-to-visible face recognition. https://doi.org/10.1109/icb2018.2018.00034 (accessed on 11th July 2025).

Jian, Z., Xie, X., Wang, L., Cao, M., & Zhang, M. (2019). Generating photographic faces from the sketch guided by attribute using GAN. *IEEE Access, 7*, 23844–23851. https://doi.org/10.1109/access.2019.2899466 (accessed on 11th July 2025).

Kazemi, H., Iranmanesh, M., Dabouei, A., Soleymani, S., & Nasrabadi, N. M. (2018). Facial attributes guided deep sketch-to-photo synthesis. https://doi.org/10.1109/wacvw.2018.00006 (accessed on 11th July 2025).

Kingma, D. P., & Ba, J. (2014). Adam: A method for stochastic optimization. https://arxiv.org/abs/1412.698 (accessed on 11th July 2025).

Lee, J., Kim, E., Lee, Y., Kim, D., Chang, J., & Choo, J. (2020). Reference-based sketch image colorization using augmented-self reference and dense semantic correspondence. https://doi.org/10.1109/cvpr42600.2020.00584 (accessed on 11th July 2025).

Li, C., & Wand, M. (2016). Combining Markov random fields and convolutional neural networks for image synthesis. https://doi.org/10.1109/cvpr.2016.272 (accessed on 11th July 2025).

Li-Liang, Z., Li, L., Wang, X., Ding, S., & Zhang, L. (2015). End-to-end photo-sketch generation via fully convolutional representation learning. https://doi.org/10.1145/2671188.2749321 (accessed on 11th July 2025).

Liu, B., Song, K., Zhu, Y., & Elgammal, A. (2021). Sketch-to-art: Synthesizing stylized art images from sketches. In *Lecture Notes in Computer Science* (pp. 207–222). https://doi.org/10.1007/978-3-030-69544-6_13 (accessed on 11th July 2025).

Liu, W., Tang, X., & Liu, J. (2007). Bayesian tensor inference for sketch-based facial photo hallucination (pp. 2141–2146). http://mmlab.ie.cuhk.edu.hk/2007/IJCAI07_hallucination.pdf (accessed on 11th July 2025).

Lu, Y., Wu, S., Tai, Y., & Tang, C. (2018). Image generation from sketch constraint using contextual GAN. In *Lecture Notes in Computer Science* (pp. 213–228). https://doi.org/10.1007/978-3-030-01270-0_13 (accessed on 11th July 2025).

Osahor, U., Kazemi, H., Dabouei, A., & Nasrabadi, N. M. (2020). Quality guided sketch-to-photo image synthesis. https://doi.org/10.1109/cvprw50498.2020.00418 (accessed on 11th July 2025).

Socher, R., Ganjoo, M., Manning, C. D., & Ng, A. Y. (2013). Zero-shot learning through cross-modal transfer. *arXiv (Cornell University), 26*, 935–943. https://arxiv.org/pdf/1301.3666 (accessed on 11th July 2025).

Wang, L., Sindagi, V. A., & Patel, V. M. (2018). High-quality facial photo-sketch synthesis using multi-adversarial networks. https://doi.org/10.1109/fg.2018.00022 (accessed on 11th July 2025).

Wang, X., & Tang, X. (2009). Face photo-sketch synthesis and recognition. *IEEE Transactions on Pattern Analysis and Machine Intelligence, 31*(11), 1955–1967. https://doi.org/10.1109/tpami.2008.222 (accessed on 11th July 2025).

Xiao, B., Gao, X., Tao, D., & Li, X. (2009). A new approach for face recognition by sketches in photos. *Signal Processing, 89*(8), 1576–1588. https://doi.org/10.1016/j.sigpro.2009.02.008 (accessed on 11th July 2025).

Yu, J., Xu, X., Gao, F., Shi, S., Wang, M., Tao, D., & Huang, Q. (2021). Toward realistic face photo–sketch synthesis via composition-aided GANs. *IEEE Transactions on Cybernetics, 51*(9), 4350–4362. https://doi.org/10.1109/tcyb.2020.2972944 (accessed on 11th July 2025).

Zhang, J., Wang, N., Gao, X., Tao, D., & Li, X. (2011). Face sketch-photo synthesis based on support vector regression. https://doi.org/10.1109/icip.2011.6115625 (accessed on 11th July 2025).

Zhang, Y., Su, G., Qi, Y., Yang, J., & Beijing University of Posts and Telecommunications. (2021). Unpaired image-to-sketch translation network for sketch synthesis. *IEEE Transactions on Image Processing*. URL (accessed on 11th July 2025).

Zhao, J., Xie, X., Wang, L., Cao, M., & Zhang, M. (2019). Generating photographic faces from the sketch guided by attribute using GAN. *IEEE Access, 7,* 23844–23851. https://doi.org/10.1109/ACCESS.2019.2899466 (accessed on 21th August 2025).

Zheng, J., Song, W., Wu, Y., Xu, R., & Liu, F. (2019). Feature encoder guided generative adversarial network for face photo-sketch synthesis. *IEEE Access, 7*, 154971–154985. https://doi.org/10.1109/access.2019.2949070 (accessed on 11th July 2025).

CHAPTER 13

Comparative Review of Image Filtering in Spatial and Frequency Domain

PRASHANT SINGH, CHITTARANJAN PRADHAN,* and PRADEEP KUMAR MALLICK

School of Computer Engineering, Kalinga Institute of Industrial Technology, Bhubaneswar, Odisha, India

*Corresponding author

ABSTRACT

Filtering is a necessary step to remove noise from a signal. The signal can be an image, video, and so on. Here, we focus on image signals. Denoising can be performed through filtering techniques in either the spatial domain or the frequency domain. This chapter presents a study of different filtering techniques and a comparative review of these techniques in both domains on noisy image data. We have used estimation parameters such as Mean Squared Error (MSE), peak signal-to-noise ratio (PSNR), and root mean square error (RMSE) for evaluation. The study provides a clear representation of each filtering technique along with its strengths.

13.1 INTRODUCTION

With advancements in different fields of Computer Science and Engineering, progress in the field of digital image processing is occurring rapidly. In the real world, signals are affected by noise during acquisition or digitization. Images become corrupted by various types of noise during transmission due to interference in the transmitting channel. Image denoising is one of the most basic problems in digital image processing and computer vision. Effective denoising techniques remove different

Intelligent Computing and Analytics in Image Processing. Samarjeet Borah, Jambi Ratna Raja Kumar, Laxmi Bewoor, Rahul Raghvendra Joshi, and Mahendra Prabhakar Deore (Eds.)

DOI: 10.1201/9781779640772-13

types of noise while preserving the important details of images (Gonzalez, Woods, and Eddins, 2017).

Linear filtering techniques used in the early development phase of image processing were key methods for image enhancement and restoration. Linear filters are easy to implement and design due to their mathematical simplicity and also possess several desirable properties. While they performed well in many applications, they did not perform effectively in the presence of additive noise or in systems involving non-linearity or Gaussian statistics. Linear filters could not efficiently denoise images corrupted with Gaussian or mixed Gaussian noise and often blurred image edges.

Linear filtering techniques also perform poorly in the presence of transient, non-stationary wide-band components. This performance degradation is due to the similarity between the signal spectrum and the noise spectrum, which causes the noise energy to spread across all coefficients in the frequency domain. In contrast, signal energy is typically concentrated in only a few coefficients. Therefore, nonlinear methods are employed, which reduce the coefficients associated with noise to zero while retaining or enlarging the coefficients representing the signal (Gonzalez, Woods, and Eddins, 2017).

Noise denoising techniques involving multi-resolution transforms generally follow three steps: linear forward transform, nonlinear thresholding, and linear inverse transform.

In this comparative study, different denoising methodologies were used for noise removal from test images while preserving image quality. Along with many advantages, these filtering algorithms also have some disadvantages. Each filtering technique differs from others in various aspects, such as computational cost, noise removal efficiency, and the quality of the resultant image after applying filters. As a result, noise removal methods can still be improved and remain an open area of research.

13.2 NOISE TYPES

Noise in images refers to random variations in various parameters such as brightness or color data present in the image. These noises are often generated by the sensing components of digital cameras or scanners. Additionally, noise can arise due to defects in film grain or unnecessary shots captured by the photon detector. Image noise is essentially a by-product of the image capture process. Different types of noise are described in subsections.

13.2.1 GAUSSIAN NOISE

Gaussian noise is also known as amplifier noise, and it is additive in nature. The primary cause of Gaussian noise in images is during the acquisition phase. This type of statistical noise is modeled using a probability density function that resembles a normal distribution (Agarwal and Kumar, 2015). It is also referred to as Gaussian distribution. The probability density function can be expressed as:

$$P(z) = \frac{1}{\sqrt{2}\pi\sigma}\pi e^{-\frac{z-\mu^2}{2\sigma^2}} \tag{1}$$

where; z is gray level; u is mean of z; σ is standard deviation; and σ^2 is variance of z.

13.2.2 SALT AND PEPPER NOISE

This type of noise operates based on the principle of invertibility—that is, it replaces dark regions with bright pixels and bright regions with dark pixels. The intensity of these pixels depends on the noise density in the image. This noise is introduced into an image due to errors during signal conversion or signal transmission. It is also called impulse noise (Kuan et al., 1985). Salt and pepper noise is expressed as:

$$P(z) = \begin{cases} P_x, forz = x \\ P_y, forz = y \\ 0, otherwise \end{cases} \tag{2}$$

For $y > x$, y appears as a light dot; otherwise, x appears as a dark dot.

13.2.3 SPECKLE NOISE

Being a type of granular noise in images, speckle noise decreases image quality in modalities such as active radar and medical ultrasound (Udomhunsakul and Wongsita, 2004). It is represented as:

$$B = A + n \times A \tag{3}$$

where; A and B are the original and introduced speckle-noised images; and n is the uniform noise image.

13.3 FILTERING TECHNIQUES

There are basically two filtering methods. Both methods have their own advantages and disadvantages and are adopted based on specific requirements, complexity, and performance.

13.3.1 SPATIAL DOMAIN FILTERING

The spatial domain represents the two-dimensional image plane. Since an image is a two-dimensional matrix of pixels, spatial domain filtering directly modifies pixel values. It can be represented as:

$$A'(x,y) = T\,[A(x,y)] \tag{4}$$

where A and A' are the original and modified images, respectively; T is the transformation function that modifies pixel values based on the neighborhood.

In spatial domain techniques, the relative position and values of neighboring pixels are critical. There are two types of spatial filters which are discussed in subsections.

13.3.1.1 Linear filter

1. **Mean Filter:** It is simple, intuitive, and used for smoothing images, that is, reducing the variation of intensities between consecutive pixels (Agarwal and Kumar, 2015). In this method, each pixel intensity is replaced by the mean value of its neighborhood, including itself. It is a type of convolution filter. The mean filter uses a kernel during the calculation of the mean. A larger window can remove noise more effectively. Although it smoothens images and eliminates unrepresented pixel values, it may cause edge blurring due to overlapping.
2. **Wiener Filter:** This linear filter is based on the MSE-optimal stationary process for images. It denoises images by addressing additive noise and blurring effects (Udomhunsakul and Wongsita, 2004). The overall Mean Squared Error (MSE) is minimized during inverse filtering and noise smoothing. While implementing spatial smoothing and managing complexity, this filter adjusts window size dynamically. It can significantly reduce noise when variance

is low and can use a large window for smoothing speckle noise. It works well when the underlying signal is smooth. However, it is difficult to perfectly restore signals affected by random noise, and the method is computationally slow.

3. **Lee Filter:** Based on the multiplicative speckle noise model, this filter uses local statistics to preserve image details (Kuan et al., 1985; Lee, 1981). It produces output by considering the average pixel intensity within a filter window. Due to its adaptive nature, the Lee filter preserves details in both low and high contrast areas and operates based on variance. While it offers better edge preservation, it struggles to remove speckle noise near edges effectively.
4. **Kuan Filter:** This filter, also based on minimum MSE, is designed for removing multiplicative noise (Kuan et al., 1987). It transforms multiplicative noise into additive noise. Similar to the Lee filter, the Kuan filter uses a different weighting function. It is considered more advanced because it avoids approximation of noise variance. The method is useful in remote sensing, oceanography, and geology, particularly for speckle noise removal. However, it depends on the ENL (effective number of looks) from the image to determine the weighting function.
5. **Frost Filter:** Based on linear convolution, the Frost filter is used for removing multiplicative noise. It is an exponentially weighted averaging filter (Frost et al., 1982). The filtering operation replaces the center pixel with a weighted sum of intensity values from its neighborhood in the kernel. The weighting factor decreases with distance from the center pixel and increases with variance. It smoothens homogeneous regions by giving higher weight to the center pixel. Although effective at removing speckle noise, it is computationally complex due to its reliance on multiple calculations.

13.3.1.2 Non-linear filter

1. **Median Filter:** By preserving the edges, this widely used filtering technique can remove noise (Lee and Kassam, 1985). It replaces the gray level of an image with the median value from a particular neighborhood. The neighborhood pattern is called the window. It is very effective at removing salt and pepper noise and is also

applicable for reducing speckle noise. It removes noise without affecting the edges. However, it is difficult to analytically evaluate the effect of a median filter, and it requires high computational cost.

2. **Max Filter:** Also known as the 100th percentile filter (Sharma and Kumar, 2016), the max filter works by determining the brightest points in the image. It divides the image into sub-image areas to find and use the maximum intensity in each sub-image area. It is best suited for reducing pepper noise.
3. **Min Filter:** Also known as the 0th percentile filter (Sharma and Kumar, 2016), this filter works completely opposite to the max filter. It finds the darkest points in the image and uses the minimum intensity in the sub-image area. It is best suited for reducing salt noise as a result of the minimum operation.

13.3.2 FREQUENCY DOMAIN FILTERING

In this method, the image is first transformed into the frequency domain. Frequency domain methods work by modifying the Fourier transform coefficients of the image (Lee, 1980; Makandar and Halalli, 2015). The term "frequency" in an image refers to the rate of change in pixel values.

1. **Ideal Low-Pass Filter:** Also known as a blurring or smoothing filter (Lee, 1980; Makandar and Halalli, 2015), it eliminates components above the cut-off frequency while allowing those below to remain unchanged. The image produced by this filter appears significantly blurrier. It affects noise more than the image data. The quality of the image can be gradually enhanced by suppressing the noise, making it suitable for identifying faint details obscured by noise.
2. **Ideal High-Pass Filter:** This filter sharpens the image (Lee, 1980; Makandar and Halalli, 2015) by highlighting fine details. However, it also amplifies noise. While it can enhance subtle image features, excessive application leads to grainy and unnatural-looking results.
3. **Butterworth Low-Pass Filter:** This filter's frequency response has no ripples in the pass band and stop band. It provides a

smoother image by gradually decreasing frequency response in the transition region. It is also known as a maximally flat filter.

4. **Butterworth High-Pass Filter:** It controls the sharpness of the filter based on its order. The resulting images are smoother than the originals, and boundaries are rarely distorted.
5. **Gaussian Low-Pass Filter:** This filter reduces the ringing effect caused by ideal low-pass filters and reduces edge content. The transition is smoother compared to the ideal filter.
6. **Gaussian High-Pass Filter:** This filter minimizes the ringing effect associated with ideal high-pass filters while enhancing edge content. Though the filtering concept remains the same, the transition is smoother.

13.4 ESTIMATION OF STATISTICAL PARAMETERS

Denoising techniques can be implemented using MATLAB. For evaluating the performance of different filtering techniques, we have considered MSE, PSNR, and RMSE.

1. **MSE (Mean Square Error):** This metric compares the pixel intensity values of the original and filtered images (Achim, Kuruoglu, and Zerubia, 2006; Makandar and Halalli, 2015). It is calculated as the mean of the squared differences between the intensities of the original and filtered images.

$$MSE = \frac{1}{AB}\sum_{i=0}^{A-1}\sum_{j=0}^{B-1}[C(i,j)-D(i,l)]^2 \tag{5}$$

where; C and D are the original and filtered images, respectively of size A rows and B columns.

2. **RMSE (Root Mean Square Error):** This is the square root of the MSE and provides an absolute difference measure between original and filtered images (Achim, Kuruoglu, and Zerubia, 2006; Samanta et al., 2022).

$$RMSE = \sqrt{MSE} \tag{6}$$

3. **PSNR (Peak Signal-to-Noise Ratio):** This is a performance measure that evaluates the quality of filtering (Achim, Kuruoglu, and Zerubia, 2006). It is defined as:

$$PSNR = 10 log_{10}((2^N - 1)^2 / MSE \quad (7)$$

13.5 RESULT ANALYSIS

For the analysis of the results, we have taken two grayscale images—namely, "Baboon.bmp" (a facial image) of size 512 × 512 and "Cameraman.png" (a distant image) of size 256 × 256. Noisy images were generated using Gaussian noise, salt and pepper noise, and speckle noise. These noisy images were then filtered using different types of spatial filters at a fixed noise density, as shown in Figures 13.1–13.3. After applying the filters, we computed the Mean Squared Error (MSE), PSNR, and root mean squared error (RMSE) values to evaluate filter performance.

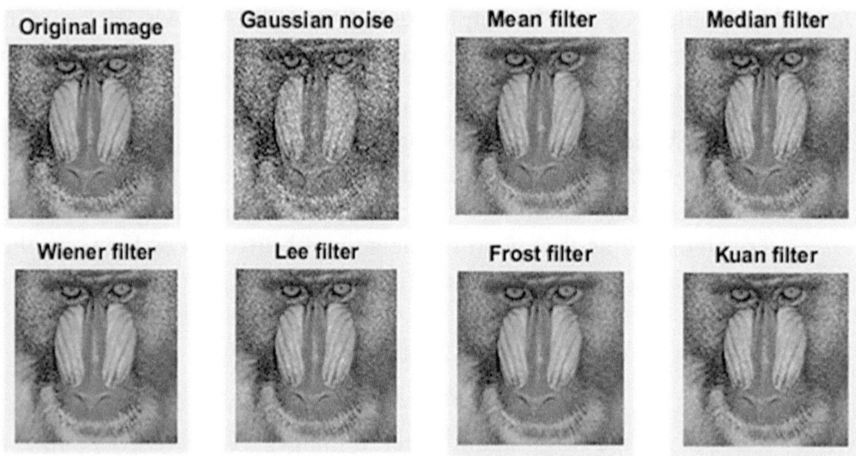

FIGURE 13.1 Images of different types of spatial filters applied on Gaussian noise.

From Table 13.1, it can be observed that the Wiener filter performs better than other filters for Gaussian noise with a mean of 0 and variance of 0.01. Additionally, the Frost and Lee filters also show good performance for Gaussian noise reduction.

TABLE 13.1 Comparison of Various Spatial Filters for Gaussian Noise (Mean = 0 and Variance = 0.01)

Filter Type	Image	MSE	PSNR	RMSE
Mean filter	Baboon	65.76	29.95	8.11
	Cameraman	47.08	31.40	6.86
Median filter	Baboon	69.08	29.74	8.31
	Cameraman	52.32	30.94	7.23
Wiener filter	Baboon	62.93	30.14	7.93
	Cameraman	40.23	32.08	6.35
Lee filter	Baboon	64.63	30.02	8.04
	Cameraman	43.58	31.74	6.60
Frost filter	Baboon	64.64	30.01	8.06
	Cameraman	42.34	31.86	6.51
Kuan filter	Baboon	65.05	29.99	8.07
	Cameraman	45.93	31.51	6.78

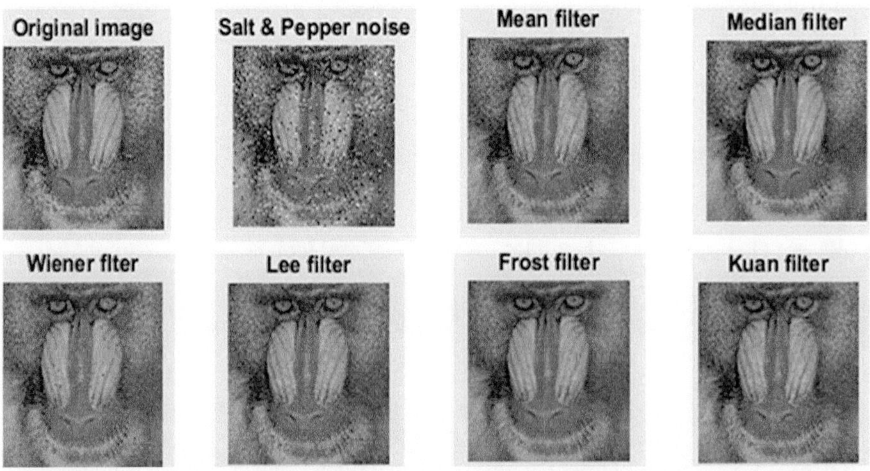

FIGURE 13.2 Images of different types of spatial filters applied on salt and pepper noise.

From Table 13.2, it has been observed that the median filter performs better than other filtering techniques for salt and pepper noise with a noise density of 0.05. The second-best performance is achieved by the Wiener filter.

TABLE 13.2 Comparison of Various Spatial Filters for Salt and Pepper Noise (Noise Density = 0.05)

Filter Type	Image	MSE	PSNR	RMSE
Mean filter	Baboon	67.27	29.85	8.20
	Cameraman	53.44	30.85	7.31
Median filter	Baboon	45.28	31.57	6.73
	Cameraman	17.95	35.59	4.24
Wiener filter	Baboon	64.11	30.06	8.01
	Cameraman	53.32	30.86	7.30
Lee filter	Baboon	66.84	29.88	8.18
	Cameraman	52.21	30.95	7.23
Frost filter	Baboon	68.69	29.76	8.23
	Cameraman	56.25	30.63	7.50
Kuan filter	Baboon	66.46	29.91	8.15
	Cameraman	52.86	30.90	7.27

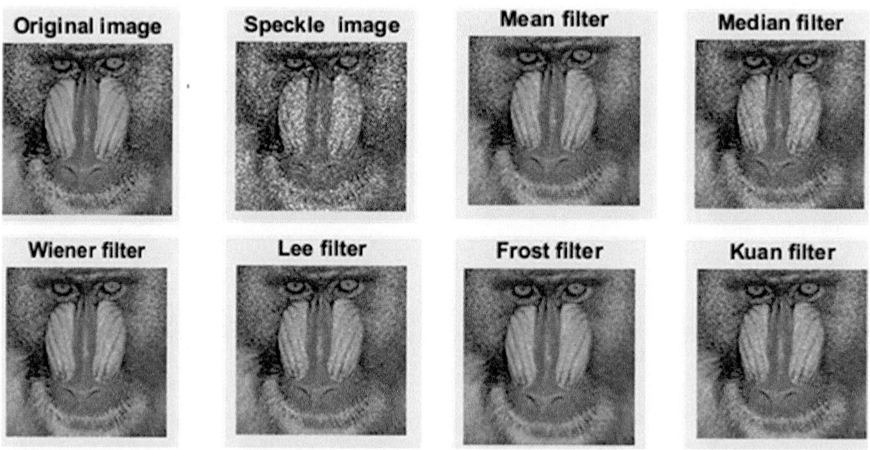

FIGURE 13.3 Images of different types of spatial filters applied on speckle noise.

For speckle noise removal, the Wiener filter performs best with a variance value of 0.04 (as shown in Table 13.3). Following the Wiener filter, the Frost filter achieves good results. It is also observed that Wiener filtering consistently performs well across all three types of noise considered in this study. Therefore, it is recommended for general-purpose noise removal.

TABLE 13.3 Comparison of Various Spatial Filters for Speckle Noise (Variance = 0.04)

Filter Type	Image	MSE	PSNR	RMSE
Mean filter	Baboon	67.27	29.85	8.20
	Cameraman	53.44	30.85	7.31
Median filter	Baboon	45.28	31.57	6.73
	Cameraman	17.95	35.59	4.24
Wiener filter	Baboon	64.11	30.06	8.01
	Cameraman	53.32	30.86	7.30
Lee filter	Baboon	66.84	29.88	8.18
	Cameraman	52.21	30.95	7.23
Frost filter	Baboon	68.69	29.76	8.23
	Cameraman	56.25	30.63	7.50
Kuan filter	Baboon	66.46	29.91	8.15
	Cameraman	52.86	30.90	7.27

The same process is followed for frequency domain filters using the same images—'Baboon.bmp' and "Cameraman.png"—of the same sizes, as shown in Figures 13.4–13.6.

FIGURE 13.4 Images of different types of frequency domain filters applied on salt and pepper noise.

From Table 13.4, it has been observed that the Gaussian low pass filter provides better results for image smoothing, whereas the Butterworth high pass filter performs better for edge detection. This study was conducted by considering an order (n) of 1, a cut-off frequency (d) of 50, and a noise density of 0.04.

TABLE 13.4 Comparison of Various Frequency Domain Filters for Salt and Pepper Noise ($n = 1$, $d = 50$, Noise Density = 0.04)

Filter Type	Image	MSE	PSNR	RMSE
Ideal low pass filter	Baboon	109.24	27.75	10.45
	Cameraman	94.33	28.38	9.71
Butterworth low pass filter	Baboon	29.32	33.46	5.41
	Cameraman	34.37	32.77	5.86
Gaussian low pass filter	Baboon	22.85	34.54	4.78
	Cameraman	30.74	33.25	5.54
Ideal high pass filter	Baboon	1227.87	17.24	35.04
	Cameraman	1411.78	16.63	37.57
Butterworth high pass filter	Baboon	968.73	18.27	31.12
	Cameraman	752.63	19.28	27.70
Gaussian high pass filter	Baboon	1046.29	17.93	32.35
	Cameraman	800.04	19.10	28.28

FIGURE 13.5 Images of different types of frequency domain filters applied on speckle noise.

In the case of speckle noise, the Butterworth low pass filter performs better for image smoothing, while the Butterworth high pass filter is efficient for edge detection (*see* Table 13.5). This study was also conducted using an order (n) of 1, a cut-off frequency (d) of 50, and a noise density of 0.04.

TABLE 13.5 Comparison of Various Frequency Domain Filters for Speckle Noise ($n = 1$, $d = 50$, Noise Density = 0.04)

Filter Type	Image	MSE	PSNR	RMSE
Ideal low pass filter	Baboon	107.75	27.81	10.38
	Cameraman	91.77	28.50	9.58
Butterworth low pass filter	Baboon	37.00	32.45	6.08
	Cameraman	43.60	31.74	6.60
Gaussian low pass filter	Baboon	33.72	32.85	5.81
	Cameraman	56.97	30.57	7.55
Ideal high pass filter	Baboon	1273.97	17.08	35.69
	Cameraman	1126.56	17.61	33.56
Butterworth high pass filter	Baboon	1015.79	18.06	31.87
	Cameraman	726.78	19.52	26.26
Gaussian high pass filter	Baboon	1086.25	17.77	32.96
	Cameraman	736.24	19.46	27.13

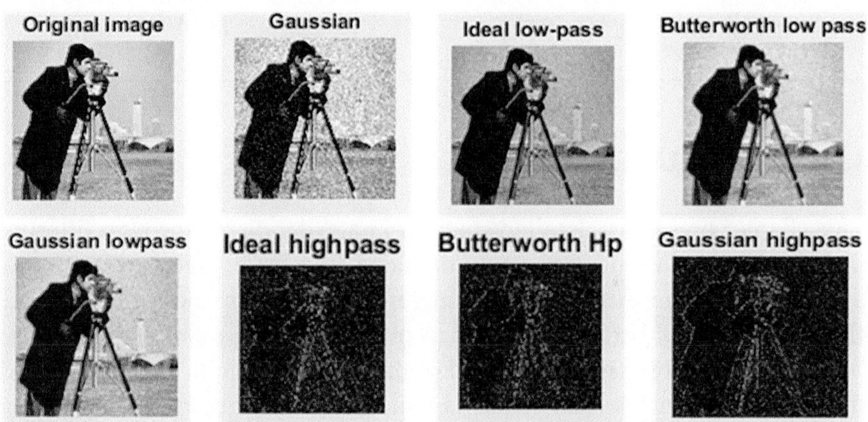

FIGURE 13.6 Images of different types of frequency domain filters applied on Gaussian noise.

Table 13.6 shows that the Gaussian low pass filter is efficient for image smoothing, whereas the Butterworth high pass filter performs better for edge detection. The parameters considered are an order (n) of 1, a cut-off frequency (d) of 50, a mean value of 0, and a variance of 0.01.

From the above results, we can conclude that the Gaussian low pass filter performs well for image smoothing, and the Butterworth high pass filter is most effective for edge detection tasks. Results using various filters are given in Table 13.6.

TABLE 13.6 Comparison of Various Frequency Domain Filters for Gaussian Noise (n = 1, d = 50, m = 0, and v = 0.01)

Filter Type	Image	MSE	PSNR	RMSE
Ideal low pass filter	Baboon	108.30	27.78	10.41
	Cameraman	90.50	28.56	9.51
Butterworth low pass filter	Baboon	30.32	33.31	5.51
	Cameraman	33.77	32.85	5.81
Gaussian low pass filter	Baboon	23.46	34.43	4.84
	Cameraman	42.64	31.83	6.53
Ideal high pass filter	Baboon	1192.62	17.37	34.53
	Cameraman	836.49	18.91	28.92
Butterworth high pass filter	Baboon	941.52	18.39	30.68
	Cameraman	539.66	20.81	23.23
Gaussian high pass filter	Baboon	1006.26	18.10	31.72
	Cameraman	526.14	20.92	22.94

13.6 CONCLUSION

This chapter presents a detailed study of image denoising through filtering in both the spatial and frequency domains. We have used MSE, PSNR, and RMSE as the evaluation parameters. The results show that Wiener filtering performs better compared to other techniques and can therefore be recommended for noise removal in the spatial domain. It is also found that the Gaussian low pass filter is suitable for image smoothing in the frequency domain, while the Butterworth high pass filter is preferable for edge detection in the frequency domain.

DECLARATION

No AI tools or any other similar tools have been used in the content creation of this chapter.

KEYWORDS

- **Butterworth low pass filter**
- **filtering**
- **frequency domain**
- **Gaussian filter**
- **linear filter**
- **noise**
- **non-linear filter**
- **spatial domain**

REFERENCES

Achim, A., Kuruoglu, E. E., & Zerubia, J. (2006). SAR image filtering based on the heavy-tailed Rayleigh model. *IEEE Transactions on Image Processing, 15*(9), 2686–2693.

Agarwal, S. K., & Kumar, P. (2015). Denoising of a mixed noise color image using new filter technique. *International Conference on Computational Intelligence and Communication Networks*, IEEE, 324–328.

Frost, V. S., Stiles, J. A., Shanmugan, K. S., & Holtzman, J. C. (1982). A model for radar images and its application to adaptive digital filtering of multiplicative noise. *IEEE Transactions on Pattern Analysis and Machine Intelligence, 4*(2), 157–166.

Gonzalez, R., Woods, R., & Eddins, S. (2017). *Digital Image Processing Using MATLAB.* McGraw Hill.

Kuan, D. T., Sawchuk, A. A., Strand, T. C., & Chavel, P. (1985). Adaptive noise smoothing filter for images with signal-dependent noise. *IEEE Transactions on Pattern Analysis and Machine Intelligence, 7*(2), 165–177.

Kuan, D. T., Sawchuk, A. A., Strand, T. C., & Chavel, P. (1987). Adaptive restoration of images with speckle. *IEEE Transactions on Acoustics, Speech, and Signal Processing, 35*(3), 373–383.

Lee, J. S. (1980). Digital image enhancement and noise filtering by use of local statistics. *IEEE Transactions on Pattern Analysis and Machine Intelligence, 2*(2), 165–168.

Lee, J. S. (1981). Refined filtering of image noise using local statistics. *Journal of Computer Graphics and Image Processing, 15*(4), 380–389.

Lee, Y., & Kassam, S. (1985). Generalized median filtering and related nonlinear filtering techniques. *IEEE Transactions on Acoustics, Speech, and Signal Processing, 33*(3), 672–683.

Makandar, A., & Halalli, B. (2015). Image enhancement techniques using high pass and low pass filters. *International Journal of Computer Applications, 109*(14), 12–15.

Samanta, S. R., Mallick, P. K., Pattnaik, P. K., Mohanty, J. R., & Polkowski, Z. (2022). *Cognitive Computing for Risk Management*. Springer.

Sharma, S., & Kumar, Er. K. (2016). A comparison of salt and pepper noise removal filters. *International Journal of Engineering Science and Computing, 6*(8), 2627–2630.

Udomhunsakul, S., & Wongsita, P. (2004). Ultrasonic speckle denoising using the combination of wavelet transform and Wiener filter. *Asian and Pacific Rim Symposium on Biophotonics*, IEEE, 195–196.

CHAPTER 14

An Analysis on Feature Descriptors Used in Computer Vision

TUKARAM UGILE[1,*] and NILESH UKE[2]

[1]*Research Scholar, Department of Computer Engineering, Vishwakarma Institute of Information Technology, Affiliated to Savitribai Phule Pune University (SPPU), Pune, Maharashtra, India*

[2]*Research Supervisor, Department of Computer Engineering, Vishwakarma Institute of Information Technology, Affiliated to Savitribai Phule Pune University (SPPU),Pune, Maharashtra, India*

*Corresponding author

ABSTRACT

This chapter presents a survey of feature descriptors used in computer vision. Recent and relevant research papers from the computer vision domain are reviewed, and a survey is conducted on various feature descriptors used in the computer vision literature for image matching, action recognition, and detection of abnormal events. A comparison is performed among these feature descriptors, their advantages and limitations are studied, and research directions are listed for their future use.

14.1 INTRODUCTION

Image matching, action recognition, and detection of abnormal events are key problems in the computer vision domain. Researchers have proposed different feature descriptors to handle these research problems. Various features are learned from input video using feature descriptors, and these features are utilized to address different research problems such as image

Intelligent Computing and Analytics in Image Processing. Samarjeet Borah, Jambi Ratna Raja Kumar, Laxmi Bewoor, Rahul Raghvendra Joshi, and Mahendra Prabhakar Deore (Eds.)

DOI: 10.1201/9781779640772-14

matching, action recognition, and detection of abnormal events. It is important to understand these feature descriptors and conduct a comparative study of them to provide research directions for researchers in the computer vision domain.

Features are important learned elements that are utilized for tasks such as image matching, action recognition, and detection of abnormal events in computer vision. Features are a small set of locations in the image that describe objects or scenes. They can be corners, edges, or symmetrical patterns in the image. Features can be extremely primitive or advanced. Initially, handcrafted and low-dimensional features were developed; later, deep learning (DL) methods and various streams and attention networks emerged to produce advanced features. These features helped perform several computer vision tasks with greater accuracy and reduced processing time. Different methods were developed by researchers to carry out various tasks in computer vision. The different feature descriptors are categorized as in Figure 14.1 based on their application.

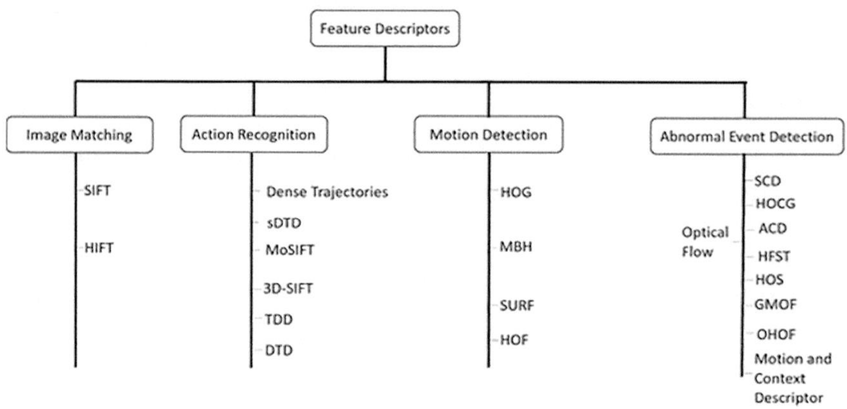

FIGURE 14.1 Feature descriptors in computer vision.

There are many feature descriptors proposed and used by researchers for various research problems in the computer vision domain. However, there has been no comprehensive survey describing all the feature descriptors, classifying them, and comparing them based on their advantages and limitations. This is a comprehensive survey that provides research directions to assist future research efforts.

This chapter is organized into different sections. Section 1 provides an introduction to the feature descriptors utilized in the computer vision domain.

Sections 2, 3, 4, and 5 provide detailed information about the feature descriptors used for image matching, action recognition, motion detection, and abnormal event detection, respectively. In each section, the respective feature descriptors, advantages, and limitations are discussed. Section 6 presents future research directions. Section 7 provides the conclusions from this review work.

14.2 FEATURE DESCRIPTORS FOR IMAGE MATCHING

Identification of some portion of an image and matching it across two or more images is called image matching. To address this research problem, researchers have proposed several methods for performing image matching tasks. Some have developed feature descriptors specifically for this task. Feature descriptors identify the discriminative points around the focus point to perform the image matching task. Listed below are some of the feature descriptors used by researchers for image matching tasks in computer vision.

14.2.1 SCALE-INVARIANT FEATURE TRANSFORM (SIFT)

Scale-invariant feature transform (SIFT) feature descriptor was suggested by David Lowe and is known for its distinctiveness. In this transform, image data is converted into coordinates corresponding to local features that are not affected by scale. Feature extraction is performed on input images, and SIFT features are stored in a database for later matching and retrieval. Euclidean distance of the feature vectors is used to determine a match between the new image and the existing image database. A subset of key points that match parameters such as object, location, scale, and orientation is considered to find the most suitable match. The SIFT feature descriptor is highly distinctive. These feature descriptors are invariant to parameters such as image location, scale, key point orientation, illumination, and noise. Scale-invariant descriptors are unaffected by scaling (Lowe, 2004).

14.2.2 HIERARCHICAL FEATURE TRANSFORMER NETWORK (HIFT)

Useful information from images is utilized, and representation ability is enhanced by using the hierarchical feature transformer network (HIFT) feature descriptor. Semantic and cross-scale-enhanced hierarchical features are used to derive dense descriptions through HIFT features. These descriptors are employed for image matching (Rao et al., 2023) (Table 14.1).

TABLE 14.1 Comparison of Image Matching Feature Descriptors

Sl. No.	Descriptor	Advantages	Limitations
1.	SIFT (Lowe, 2004; Sun et al., 2009)	Invariant to scaling, rotation, image location, occlusion, orientation of key points, and so on. Unaffected by changes in illumination, 3D camera viewpoint (Lowe, 2004).	SIFT descriptors use only monochrome intensity images (Lowe, 2004).
2.	Hierarchical feature transformer network (HIFT) (Rao et al., 2023)	Representativeness and hierarchical clues in the images are utilized (Rao et al., 2023).	Cannot work well under textureless scenarios (Rao et al., 2023).

14.3 FEATURE DESCRIPTORS FOR ACTION RECOGNITION

Recognizing actions from videos is one of the foremost research problems. Actions recognized from videos are dependent on the relationship within the spatio-temporal information. To recognize actions effectively from video, it is important to understand this spatio-temporal relationship. This relationship is effectively captured by the feature descriptors proposed by various researchers for action recognition. A few of these descriptors are discussed here.

14.3.1 DENSE TRAJECTORIES

Effective representation of videos is achieved using dense trajectories. Sampling of dense points is performed from each frame, and tracking is based on displacement information using a dense optical flow field. Camera motion is canceled from the optical flow using the estimated camera motion (Wang and Schmid, 2013).

14.3.2 SEQUENTIAL DEEP TRAJECTORY DESCRIPTOR (SDTD)

Detection of long-term motions was achieved by the sequential deep trajectory descriptor (sDTD), proposed by Shi et al. They projected the dense trajectories onto a two-dimensional plane, and effective representation was learned using a CNN-RNN network. Long-term and short-term motions, as well as static spatial features, were captured using three different streams (Shi et al., 2017).

14.3.3 MOSIFT DESCRIPTORS

The MoSIFT feature descriptor was proposed using motion-related and appearance-related information by Chen et al. Optical flow captures the motion information, and it is integrated with the appearance-related information. However, the major disadvantage of this descriptor is that it is computationally expensive. Real-time detection is not promising when using this descriptor for abnormal events (Chen et al., 2010).

14.3.4 3D-SIFT DESCRIPTOR

Representation of videos was achieved using the Bag of Words approach as proposed by Paul et al. using the three-dimensional SIFT descriptor. The 3D videos are effectively represented using this descriptor for action recognition applications. The spatio-temporal relationship is accurately captured by this descriptor (Scovanner, Ali, and Shah, 2007).

14.3.5 TRAJECTORY-POOLED DEEP CONVOLUTIONAL DESCRIPTOR (TDD)

Advantages from hand-crafted and deeply learned features are combined in the TDD descriptor proposed by Limin et al. DL-based architectures are used to learn convolutional feature maps that are discriminative in nature, followed by sampling and aggregation of the convolutional features into TDDs using pooling (Wang, Qiao, and Tang, 2015).

14.3.6 DEEP TRAJECTORY DESCRIPTOR (DTD)

Shi et al. proposed a new descriptor called the deep trajectory descriptor (DTD) (Shi et al., 2017). In this descriptor, dense trajectories were extracted from the input video frames and converted into two-dimensional planes. Dense trajectories were effectively learned using CNN (Shi et al., 2017) (Table 14.2).

TABLE 14.2 Comparison of Action Recognition Feature Descriptors

Sl. No.	Descriptor	Advantages	Limitations
1.	Dense trajectories-based approach (Wang and Schmid, 2013)	Motion information is captured well. Irregular motions and shot boundaries are captured effectively. The method provides a robust and efficient representation (Wang and Schmid, 2013).	Motion blur affects performance (Wang and Schmid, 2013). Long-term motion actions are not considered (Shi et al., 2017).
2.	Sequential deep trajectory descriptor (Shi et al., 2017)	Static spatial motion and both short-term and long-term motion are detected using three-stream networks (Shi et al., 2017).	Performance can be improved by focusing on key frames using attention mechanisms (Shi et al., 2017).
3.	MoSIFT descriptors (Chen et al., 2010)	Uses appearance and motion information. Improved accuracy (Chen et al., 2010).	Computationally expensive and not suitable for real-time applications (Chen et al., 2010).
4.	3D-SIFT (Scovanner, Ali, and Shah, 2007)	Better representation for 3D videos and improved performance compared to 2D descriptors (Scovanner, Ali, and Shah, 2007).	Recognition of actions performed at different rates is a future area for improvement (Scovanner, Ali, and Shah, 2007).
5.	TDD (Wang, Qiao, and Tang, 2015)	Combines advantages of deep learning-based and hand-crafted features (Wang, Qiao, and Tang, 2015).	–
6.	DTD (Shi et al., 2017)	Compact and powerful representation of dense trajectories (Shi et al., 2017).	Due to high overlap of dense trajectories, it is difficult to learn a good representation (Shi et al., 2017).

14.4 FEATURE DESCRIPTORS FOR MOTION DETECTION

14.4.1 HOG DESCRIPTOR

Motion information is gathered from every spatial point using the histogram of oriented gradients (HOG) descriptor, and the bag of features (BoF) is utilized for encoding. HOG descriptors provide better performance compared to wavelets. This is a simple but powerful dense image descriptor.

14.4.2 MOTION BOUNDARY HISTOGRAM (MBH) DESCRIPTORS

Local motion edge orientations are captured using motion boundary histogram (MBH) descriptors, as proposed by Dalal et al. (2006) Two flow components are considered, and their local gradients are taken separately; the respective gradient magnitudes and orientations are then identified. MBH descriptors work well with camera motion. Camera motion is canceled out as optical flow gradients are encoded in these descriptors, making them effective local descriptors (Dalal, Triggs, and Schmid, 2006).

14.4.3 SPEEDED UP ROBUST FEATURES (SURF)

The speeded up robust features (SURF) descriptor uses a Gaussian second derivative mask and relies on local Haar wavelet responses. It is similar to the SIFT feature descriptor but performs better in terms of accuracy and speed. SURF is a scale- and rotation-invariant descriptor.

14.4.4 HOF DESCRIPTOR

Local motion information is encoded using optical flow fields in histograms of optical flow (HOF) descriptors (Table 14.3).

TABLE 14.3 Comparison of Motion Detection Feature Descriptors

Sl. No.	Feature Descriptor	Advantages	Limitations
1.	HOG descriptor (Dalal and Triggs, 2005)	Produces better results compared to SIFT descriptors and fewer false positives than Haar wavelet-based detectors (Dalal and Triggs, 2005). Provides good representation of human contours (Kakade, Uke, and Jain, 2014).	Captures only static information. Context-based abnormal events cannot be detected.
2.	MBH descriptors (Dalal, Triggs, and Schmid, 2006)	More discriminative and robust to camera motion. Good representation of local information (Lo and Tsoi, 2015).	Underperforms when motion and appearance features are combined (Dalal, Triggs, and Schmid, 2006).
3.	SURF (Bay et al., 2008)	Offers better accuracy and speed than SIFT (Bay et al., 2008). Robust to motion blur (Wang and Schmid, 2013). Fast, rotation- and scale-invariant. Good repeatability, distinctiveness, and robustness (Bay et al., 2008).	Does not perform well with rotations.
4.	HOF (Dalal, Triggs, and Schmid, 2006)	Provides the best performance for moving objects.	Performance decreases in the presence of camera motion.

14.5 FEATURE DESCRIPTORS FOR ABNORMAL EVENT DETECTION

14.5.1 STRUCTURAL CONTEXT DESCRIPTOR (SCD)

The visual structural context of individuals is captured using this feature descriptor. Relationships among individuals in a crowd are modeled using the structural context descriptor (SCD). A multi-object tracker is associated with different targets across different frames. Analysis of the variation of SCD is performed for detecting abnormalities in the crowd (Yuan, Fang, and Wang, 2015).

14.5.2 HOCG DESCRIPTOR (HISTOGRAM OF ORIENTED CONTEXTUAL GRADIENT)

HOG-based gradients have clear disadvantages in that they are unable to detect events that are abnormal in certain contexts. Understanding the context of events is essential for detecting context-based abnormal events. Therefore, context-based descriptors were proposed to learn the contextual composition of events. The HOCG descriptor is one such descriptor. It helps in understanding how the context of the event is composed. HOCG descriptors outperform HOG descriptors in terms of performance (Hu et al., 2018).

14.5.3 ACTION CONTEXT DESCRIPTOR (ACD)

Tian et al. demonstrated that context-based descriptors are effective for detecting abnormal events. They modeled the context information between persons and among groups. Three methods were used to model the context: one at the structure level, another at the feature level, and a third approach that combined the previous two methods (Lan et al., 2012).

14.5.4 HFST DESCRIPTORS (HIGH-FREQUENCY AND SPATIOTEMPORAL TRANSFORM)

Detection of abnormal events was carried out using high-frequency and spatiotemporal (HFST) features by Wang et al. (2012). They applied this feature descriptor for both local and global crowd behavior detection. Dynamic characteristics from the scene in the time dimension were analyzed. Local and global scenes were modeled using linear discriminant analysis (LDA) and multiple hidden Markov models (HMMs). High-frequency information is more prevalent in abnormal regions compared to normal regions of the image. This descriptor is useful for distinguishing between abnormal and normal events (Wang et al., 2012).

14.5.5 HOS DESCRIPTORS (HISTOGRAM OF ORIENTED SWARMS)

The dynamics of crowded environments were captured by Kaltsa et al. using a feature descriptor called the histogram of oriented swarms (HOS). This descriptor leverages the intelligence of swarms to capture motion dynamics, along with appearance features, for effective representation of the scene (Kaltsa et al., 2015).

14.5.6 GMOF AND OHOF DESCRIPTORS

The Gaussian model of optical flow (GMOF) was proposed by Zhang et al. to extract violent regions. To differentiate between violent and non-violent events, they also proposed the OHOF descriptor (orientation histogram of optical flow) (Zhang et al., 2016).

14.5.7 MOTION AND CONTEXT FEATURE DESCRIPTORS

Contextual information related to motion and spatial dimensions was used by Zhu et al. to build event context for abnormal event detection. For a specific activity class, frequent spatial and motion contexts were learned and used to detect abnormal events. Any contextual patterns deviating from the learned patterns were treated as abnormal. Detection of point, contextual, and collective anomalies is performed using motion and context information in this descriptor (Zhu, Nayak, and Roy-Chowdhury, 2013) (Table 14.4).

TABLE 14.4 Comparison of Abnormal Event Detection Feature Descriptors

Sl. No.	Feature Descriptor	Advantages	Limitations
1.	Structural context descriptor (Yuan, Fang, and Wang, 2015)	Abnormalities can be detected using structural context descriptor (SCD) in crowded environments (Yuan, Fang, and Wang, 2015), using target objects from the crowd.	This descriptor faces difficulties in severe weather conditions (Yuan, Fang, and Wang, 2015), such as foggy and rainy days (Yuan, Fang, and Wang, 2015).
2.	Histogram of oriented contextual gradient descriptor (Hu et al., 2018)	Captures event's contextual information (Hu et al., 2018). Performs better compared to pixel-based descriptors.	HOCG descriptors remain unexplored for action recognition and crowd activity recognition.
3.	Action context descriptor (Lan et al., 2012)	Effective due to the use of context information (Lan et al., 2012).	Activities of multiple groups (Lan et al., 2012) occurring simultaneously in a scene are not considered. Complex structures and temporal dependencies among actions are also not addressed.
4.	Motion and context feature descriptors (Zhu, Nayak, and Roy-Chowdhury, 2013)	Utilizes spatial and motion context. Recognition accuracy is improved (Zhu, Nayak, and Roy-Chowdhury, 2013).	–
5.	High-frequency and spatiotemporal (HFST) descriptors (Wang et al., 2012)	Good detection accuracy and efficiency. Captures dynamic characteristics of the scene (Wang et al., 2012).	–
6.	Histogram of oriented swarms (HOS) descriptors (Kaltsa et al., 2015)	High detection rate; performs well under occlusions, local noise, and scale variations. Lower computational cost (Kaltsa et al., 2015).	–
7.	Gaussian model of optical flow (GMOF) (Zhang et al., 2016)	Good detection accuracy and processing speed. Fast and robust in complex backgrounds (Zhang et al., 2016).	Limitations exist for flat and homogeneous surfaces.
8.	Orientation histogram of optical flow (OHOF) (Zhang et al., 2016)	High detection accuracy and processing speed. Scale- and rotation-invariant; supports real-time performance (Zhang et al., 2016).	Sensitive to illumination variation and view angle changes. More false alarms when target contrast is low.
9.	Optical flow (Mehran, Moore, and Shah, 2010)	Records motion between two pixels across two frames (Mehran, Moore, and Shah, 2010).	Cannot capture long-range temporal dependencies (Mehran, Moore, and Shah, 2010). Contextual information is also not captured.

14.6 FUTURE RESEARCH DIRECTIONS

Future descriptors need to possess the properties discussed in subsections.

14.6.1 DISTINCTIVE

Feature descriptors need to be distinctive so that they serve their intended purpose effectively. If descriptors are more distinctive, two views of a scene or object can be matched more easily. There is a higher probability of matching distinctive features from the feature database (Leo et al., 2008; Lowe, 2004).

14.6.2 INVARIANCE TO ROTATION, SCALE, AND LOCATION OF FEATURE

Feature descriptors should be invariant to scaling, rotation, and the location of features. Their performance should remain unaffected by these parameters.

14.6.3 ACCURACY

Accuracy is a critical requirement when designing feature descriptors.

14.6.4 PERFORMANCE

Future descriptors should deliver strong performance for abnormal event detection to be applicable in real-time scenarios.

14.6.5 LESS FALSE POSITIVES (FPS)

Feature descriptors should yield fewer false positives (FPs) in their results.

14.6.6 CONTEXT INFORMATION

Future descriptors should be designed to incorporate the contextual information of events and detect context-based features effectively.

14.7 CONCLUSION

A survey of different feature descriptors used for image matching, object recognition, and abnormal event detection is presented, and research directions are provided for future work. Future descriptors should be more distinctive and capable of supporting real-time applications. Their performance and accuracy are critical.

DECLARATION

No AI tools or any other similar tools have been used in the content creation of this chapter.

KEYWORDS

- **abnormal event detection**
- **action recognition**
- **computer vision**
- **feature descriptors**
- **image matching**
- **optical flow**
- **orientation histogram**
- **real-time applications**

REFERENCES

Bay, H., Ess, A., Tuytelaars, T., & Van Gool, L. (2008). Speeded-up robust features (SURF). *Computer Vision and Image Understanding, 110*(3), 346–359.

Chen, M., Mummert, L., Pillai, P., Hauptmann, A., & Sukthankar, R. (2010). Exploiting multi-level parallelism for low-latency activity recognition in streaming video. In *Proceedings of the First Annual ACM SIGMM Conference on Multimedia Systems* (pp. 1–12). Phoenix, Arizona, USA: ACM. https://doi.org/10.1145/1730836.1730838 (accessed on 11th July 2025).

Dalal, N., & Triggs, B. (2005). Histograms of oriented gradients for human detection. In *2005 IEEE Computer Society Conference on Computer Vision and Pattern Recognition (CVPR'05), 1* (pp. 886–893). IEEE.

Dalal, N., Triggs, B., & Schmid, C. (2006). Human detection using oriented histograms of flow and appearance. In *Computer Vision–ECCV 2006: 9th European Conference on Computer Vision, Graz, Austria, May 7–13, 2006. Proceedings, Part II, 9* (pp. 428–441). Springer Berlin Heidelberg.

Hu, X., Huang, Y., Duan, Q., Ci, W., Dai, J., & Yang, H. (2018). Abnormal event detection in crowded scenes using histogram of oriented contextual gradient descriptor. *EURASIP Journal on Advances in Signal Processing, 2018*(1), 54. https://doi.org/10.1186/s13634-018-0574-4 (accessed on 11th July 2025).

Kakade, S., Uke, N. J., & Jain, N. (2014). Real-time human detection using HOG feature as human descriptor and variable size sliding window. *Advances in Computer Science and Information Technology, 1*(1), 3.

Kaltsa, V., Briassouli, A., Kompatsiaris, I., Hadjileontiadis, L. J., & Strintzis, M. G. (2015). Swarm intelligence for detecting interesting events in crowded environments. *IEEE Transactions on Image Processing, 24*(7), 2153–2166. https://doi.org/10.1109/TIP.2015.2409559 (accessed on 11th July 2025).

Lan, T., Wang, Y., Yang, W., Robinovitch, S. N., & Mori, G. (2012). Discriminative latent models for recognizing contextual group activities. *IEEE Transactions on Pattern Analysis and Machine Intelligence, 34*(8), 1549–1562. https://doi.org/10.1109/TPAMI.2011.228 (accessed on 11th July 2025).

Leo, M., D'Orazio, T., Spagnolo, P., Mazzeo, P. L., & Distante, A. (2008). SIFT-based ball recognition in soccer images. In Elmoataz, A., Lezoray, O., Nouboud, F., & Mammass, D. (Eds.), *Image and Signal Processing. ICISP 2008. Lecture Notes in Computer Science, 5099*. Springer, Berlin, Heidelberg. https://doi.org/10.1007/978-3-540-69905-7_30 (accessed on 11th July 2025).

Lo, S. L., & Tsoi, A. C. (2015). Motion boundary trajectory for human action recognition. In *Computer Vision–ACCV 2014 Workshops: Singapore, Singapore, November 1–2, 2014, Revised Selected Papers, Part I, 12* (pp. 85–98). Springer International Publishing.

Lowe, D. G. (2004). Distinctive image features from scale-invariant keypoints. *International Journal of Computer Vision, 60*(2), 91–110. https://doi.org/10.1023/B:VISI.0000029664.99615.94 (accessed on 11th July 2025).

Mehran, R., Moore, B. E., & Shah, M. (2010). A streak line representation of flow in crowded scenes. In *Computer Vision–ECCV 2010: 11th European Conference on Computer Vision, Heraklion, Crete, Greece, September 5–11, 2010, Proceedings, Part III, 11* (pp. 439–452). Springer Berlin Heidelberg.

Rao, Y., Ju, Y., Wang, S., Gao, F., Fan, H., & Dong, J. (2023). Learning enriched feature descriptor for image matching and visual measurement. *IEEE Transactions on Instrumentation and Measurement, 72*, Art. No. 5008512, 1–12. https://doi.org/10.1109/TIM.2023.3249237 (accessed on 11th July 2025).

Scovanner, P., Ali, S., & Shah, M. (2007). A 3-dimensional SIFT descriptor and its application to action recognition. In *Proceedings of the 15th ACM International Conference on Multimedia* (pp. 357–360).

Shi, Y., Tian, Y., Wang, Y., & Huang, T. (2017). Sequential deep trajectory descriptor for action recognition with three-stream CNN. *IEEE Transactions on Multimedia, 19*(7), 1510–1520. https://doi.org/10.1109/TMM.2017.2666540 (accessed on 11th July 2025).

Sun, J., Wu, X., Yan, S., Cheong, L. F., Chua, T. S., & Li, J. (2009). Hierarchical spatio-temporal context modeling for action recognition. In *2009 IEEE Conference on Computer Vision and Pattern Recognition* (pp. 2004–2011). IEEE.

Wang, B., Ye, M., Li, X., Zhao, F., & Ding, J. (2012). Abnormal crowd behavior detection using high-frequency and spatio-temporal features. *Machine Vision and Applications, 23*(3), 501–511. https://doi.org/10.1007/s00138-011-0341-0 (accessed on 11th July 2025).

Wang, H., & Schmid, C. (2013). Action recognition with improved trajectories. In *Proceedings of the IEEE International Conference on Computer Vision* (pp. 3551–3558).

Wang, L., Qiao, Y., & Tang, X. (2015). Action recognition with trajectory-pooled deep-convolutional descriptors. In *Proceedings of the IEEE Conference on Computer Vision and Pattern Recognition* (pp. 4305–4314).

Yuan, Y., Fang, J., & Wang, Q. (2015). Online anomaly detection in crowd scenes via structure analysis. *IEEE Transactions on Cybernetics, 45*(3), 548–561. https://doi.org/10.1109/TCYB.2014.2330853 (accessed on 11th July 2025).

Zhang, T., Yang, Z., Jia, W., Yang, B., Yang, J., & He, X. (2016). A new method for violence detection in surveillance scenes. *Multimedia Tools and Applications, 75*(12), 7327–7349. https://doi.org/10.1007/s11042-015-2648-8 (accessed on 11th July 2025).

Zhu, Y., Nayak, N. M., & Roy-Chowdhury, A. K. (2013). Context-aware activity recognition and anomaly detection in video. *IEEE Journal of Selected Topics in Signal Processing, 7*(1), 91–101. https://doi.org/10.1109/JSTSP.2012.2234722 (accessed on 11th July 2025).

CHAPTER 15

Monitoring of Human Pose for Tracking Gym Activities

AJITKUMAR SURESHRAO SHITOLE, MAHESH GAIKWAD,*
PRANAV GAIKWAD, PRATHAMESH BHISE, and YASH DUSANE

Department of Computer Engineering, International Institute of Information Technology, Pune, Maharashtra, India

*Corresponding author

ABSTRACT

This research work introduces a technology-integrated fitness system aimed at enhancing exercise effectiveness and safety. The proposed platform ensures users perform exercises like push-ups and squats with correct posture, reducing injury risks. By recognizing proper form, users can exercise more accurately and confidently. The system features a user-friendly interface displaying schedules and repetition counts, with future potential for real-time posture feedback. Using advanced camera-based posture estimation and machine learning, it provides instant alerts when a user's form deviates, promoting safe and efficient workouts. The platform also supports personalized tracking and performance analysis. Ultimately, this technology aims to revolutionize fitness by offering guided, interactive workout experiences that improve both technique and health outcomes.

15.1 INTRODUCTION

Human pose estimation is widely used today in several fields, including action detection, sports analysis, and human–computer interfaces (Bertugl et al., 2021; Kolotouros et al., 2019; Wang et al., 2021). This paper explores the application of human pose estimation. Using images or

Intelligent Computing and Analytics in Image Processing. Samarjeet Borah, Jambi Ratna Raja Kumar, Laxmi Bewoor, Rahul Raghvendra Joshi, and Mahendra Prabhakar Deore (Eds.)

DOI: 10.1201/9781779640772-15

videos, the study addresses the challenging task of estimating the three-dimensional (3D) joint positions of the human body. In computer vision, 3D human pose estimation is gaining increasing attention, with applications in robotics, human motion analysis, and human–computer interaction. However, this task remains difficult due to the lack of in-the-wild datasets and the uncertainty in depth estimation. Over the past decade, numerous techniques have evolved, many of which employ deep learning (DL). This study is distinct in that it offers a thorough literature analysis focused on DL-based approaches. It provides a comprehensive review of the field (Schlett, Rathgeb, and Busch, 2021; Wang et al., 2021) by analyzing widely used benchmark datasets and discussing the advantages and limitations of various methods for tracking gym users' activity and providing a safe training experience. Our aim is to reshape the perception of fitness in a rapidly evolving technological landscape by promoting a holistic approach to health and wellness, rather than relying solely on routines (Liuhao et al., 2019).

The proposed solution is not merely software—it is a companion, a virtual gym buddy designed to enhance, engage, and personalize workouts. Technological innovations such as the human pose detection system (Alam et al., 2022) are revolutionizing the health and fitness sector and improving training outcomes. This device transforms gym activities by accurately capturing the locations and motions of key body points, including the head, shoulders, elbows, hips, knees, and feet. It also provides comprehensive posture and activity analysis.

As technology advances, human pose estimation has established itself as a standard in action recognition, sports analysis (Alam et al., 2022), and human–computer interfaces. Our study builds on this technology by monitoring gym activities to ensure a secure and injury-free workout. We aim to revolutionize workout practices in this fast-paced technological age by placing an emphasis on overall health and well-being. Our system is more than just software; it acts as a virtual personal trainer, customizing workout plans to meet the needs of each user, tracking repetitions, ensuring proper technique, and encouraging peak performance—all from the comfort of their own home.

Given the challenges of daily life, the difficulty of scheduling exercise, and the need for motivation, our approach is remarkably human-centric (Cai et al., 2019). This software functions as a virtual personal trainer that recognizes and celebrates every achievement, going beyond simple repetition counting. Our goal is to provide a space where individuals can unite to celebrate one another's successes and transform personal health goals into shared aspirations.

With the use of cutting-edge technology, personalized guidance, and empathetic understanding, we completely transform the fitness journey. It is not only about building muscle; it is about becoming a stronger, more confident, and resilient version of yourself. In this proposed system—where fitness and individual greatness meet—every workout is respected. Every session is a step in the right direction toward wellness. Together, let's succeed, laugh, and sweat (Shitole and Devare, 2019).

15.2 RELATED WORK

The implementation of self-tracking is essential for the advancement of 3D human projection. This work addresses the task of improving 3D human estimation accuracy without relying on labor- and capital-intensive 3D ground truths. The authors introduce EpipolarPose, a self-tracking system that does not require external camera data or 3D ground truth data. Instead, it uses multiple viewports to estimate 2D poses during training and then applies epipolar geometry to determine 3D joint positions and camera geometry. This data is subsequently used to train 3D pose estimation. EpipolarPose demonstrates state-of-the-art performance compared to weak/self-tracking algorithms on widely used datasets such as Human3.6M and MPI-INF-3DHP. Furthermore, this work contributes to 3D self-tracking human estimation by introducing a new performance metric, called pose structure score (PSS), to evaluate the model's performance against ground truth data (Kocabas, Karagoz, and Akbas, 2019; Li et al., 2021). In particular, our method is inspired by the effective use of multi-view geometry to achieve accurate 3D human estimation without a heavy reliance on 3D ground truth data (Wand and Rosenhahn, 2019).

The analysis of individual bodies and their spatial relationships is a common method for estimating human pose. This can be illustrated using an image. Convolutional neural networks (CNNs) are used to detect real body postures in human poses. However, complexity increases when multiple people are involved. A popular system first identifies individuals and then automatically estimates their actions. Some approaches incorporate interpersonal relationships by considering shading, symmetry, and long-term dependencies.

Part affinity fields (PAF) is a recent technology that enables multi-person pose estimation by encoding relationships between body parts. PAF simplifies the process and improves prediction accuracy compared to earlier methods. Enhancements to PAF have made models faster and more accurate. Integrating multiple search techniques also reduces processing time and increases precision.

This method has been integrated into the OpenPose open-source project, which can detect key points of the body, feet, hands, and face. It is a valuable tool across many applications, as it ensures accuracy and supports robust predictions.

Understanding how hands interact with objects is complex and requires detailed analysis. One approach uses RGB or RGB-D data to track hand position and movement. Another specialized method includes both hand and object modeling, recognizing that object shape influences hand orientation. Previous methods often required large storage capacity and expensive equipment, which limited their usability in capturing fine hand-object interactions.

We demonstrate how RGB images can be used to extract object information and track interactions, enabling automatic data collection for deeper analysis. Our approach employs 3D voxel meshes and a technique that transforms images into a reduced coordinate space to estimate object geometry from RGB input. By combining 2D and 3D samples in the same space, we can ensure a tighter match and higher accuracy.

Estimating 6D object pose introduces additional challenges, particularly in determining whether an object is symmetrical or has defined orientation. CNNs are used in high-resolution applications to estimate object position and orientation and also address conflicting viewpoints. When new objects are introduced, synthetic data is often used for training. However, synthetic and real data are not always equivalent. Some approaches address this using domain randomization and photorealistic rendering to make synthetic data appear more realistic. Our approach differs: we use a GAN, an AI technique capable of transforming artificial object data into realistic representations without human input. This improves model training effectiveness.

These methods contribute to better understanding of hand-object interactions, improving object prediction models for real-world applications.

This aligns directly with our objective of enabling broad, easy access to fitness technologies. Furthermore, spatial–temporal correlations for

3D pose estimation using graph convolutional networks offer key techniques for enhancing real-time depth information accuracy (Cao et al., 2019; Kokic, Kragic, and Bohg, 2019), and provide additional insight into modeling human emotions.

Here, we examine 3D human pose estimation—the process of determining a person's position in 3D space using a 2D image. Researchers continue to explore new solutions to challenges in this area. Early models relied on datasets such as Human3.6M and HumanEva; however, these models face difficulties with real-world imagery because the original data was collected in controlled laboratory environments.

To address this limitation, a two-stage approach was introduced: the model first estimates the 2D pose and then infers the 3D position. Some methods incorporate multi-task learning and passive tracking by leveraging both 2D and 3D data. Nevertheless, these approaches still struggle with uncertainty and a lack of consistency across varied environments.

Another well-researched strategy for 3D pose estimation involves using modified human pose profiles. While synthetic data is widely used in computer vision, transferring knowledge from synthetic to real data remains problematic, particularly for estimating human pose.

A promising direction is to refine domain adaptation techniques, which help bridge the gap between synthetic and real data by preserving consistency and structure. Earlier methods often ignored the nuances of human anatomical statistics in favor of task segmentation and image integration.

In this study, we address the challenge of interpreting synthetic and real 3D human pose profiles. Our proposed method, called adaptive human pose (AHuP), incorporates novel domain adaptation techniques within both feature space and pose space. To mitigate the negative effects of domain shifts on 3D estimation, AHuP maximizes the correlation between human pose predictions. The system includes optimization strategies and bone-based modeling to improve the accuracy and robustness of 3D human body estimation—without relying directly on labeled 3D body data.

Research in mesh estimation (also known as mesh registration) began in the 1990s. With the advent of RGB-D sensors, the ability to reconstruct 3D meshes using depth or RGB-D data improved significantly. More recently, attention has shifted to RGB-based methods, and synthetic image generation has been explored to overcome the shortage of annotated 3D data. While our approach adopts a similar architectural foundation, it also incorporates connectivity modeling. Unlike methods that focus solely on skeletal data, we aim to generate a complete hand mesh and develop a deeper understanding of structural connectivity.

A popular research area is the identification of three-dimensional objects using the CNN framework. Many methods have been explored, such as mesh surfaces, point clouds, and voxels. We use mesh surfaces because they can effectively model structural and physical interactions. Compared to existing methods that typically operate with simple CAD models, our method incorporates a rod structure and utilizes more complex models.

Previous studies have applied various RGB or RGB-D visualization techniques along with optimization or classification methods for 3D reconstruction. While most of these approaches involve complex object structures, our solution differs by using an end-to-end learning CNN algorithm that leverages unique hand patterns and physical constraints, achieving high accuracy and performance.

We ensure that our proposed method not only estimates body posture accurately but also draws inspiration from existing methods to analyze body movement during the observation period (Liu, Sehgal, and Ostadabbas, 2022; Yana et al., 2019).

An important part of our application's prediction and user tracking is a series of predictions for the future. The use of continuous variable recurrent neural networks (C-VRNNs) and their combination with tracking techniques to improve instant prediction has a significant impact on our method. Our work based on the investigation of deep falls is important for safety in identifying inappropriate behaviors and preventing injuries. By implementing explainability, we can analyze the user's physical condition during operation and provide instant feedback. The method proposed in this paper provides a new approach for 3D human estimation in movies by separating the position into two main components: bone orientation estimation and bone length estimation. This method provides more precise and accurate information about the anatomy of the body by focusing on specific bones in the human skeleton. This study also demonstrates the importance of physical disparity in video sequences, which improves the use of this technique in different time frames. Experimental results confirm the effectiveness of the method and demonstrate its potential applications in human body analysis and computer vision, including perception, motion, and human–computer interaction. As stated by Shitole and Devare (2018), the core of our system lies in shape. We adapt similar methods to generate accurate estimates. Our body's ability to track multiple joints in real time is inspired by OpenPose's multi-person 2D projection design. Our body's ability to predict what is happening on film is directly affected by the use of temporal consistency for 3D prediction, which is essential for immediate feedback.

15.3 PROPOSED ARCHITECTURE AND METHODOLOGY

The ingeniously crafted proposed solution blends state-of-the-art technology with user-centric design principles. The foundation of our solution is a well-structured architecture that ensures user satisfaction and seamless operation. The key component of our application is its robust system design. ReactJS was used to construct an intuitive front end for our system (Chen et al., 2021). This interface is easy to navigate, allowing anyone to monitor their fitness. The backend, which securely and efficiently handles user data, is powered by Flask. Massive volumes of user data are effectively stored and organized using our reliable database, MongoDB. FastAPI acts as a bridge, facilitating smooth communication between different components.

The system architecture diagram clearly illustrates the various layers and complex components of a fully functional gym pose tracking website. At the forefront is the aesthetically pleasing user interface, designed to engage and interact with users. The API layer is essential for enabling data transfer and communication between the presentation layer and the rest of the system. The middle layer performs the critical function of collecting data from multiple sources, such as joints and visual inputs, which is essential for accurate pose tracking. The system's ability to compute angles and identify landmarks enhances its accuracy in tracking user poses. Flask, a powerful web framework, manages web sessions and serves as a mediator between various layers. The system's efficiency is further improved through the integration of additional libraries, such as OpenCV (Figure 15.1).

Basic illustrations, including system diagrams, state diagrams, and functional diagrams, are used throughout the text to enhance reader understanding. This structured approach allows readers to clearly grasp the explanations and visualize the information presented.

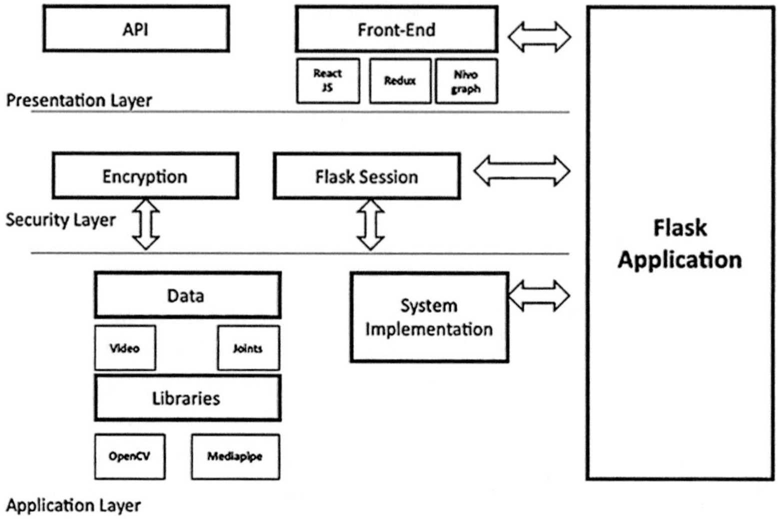

FIGURE 15.1 System architecture.

Technical precision and user support are central to our approach. The system utilizes the MediaPipe library (Javeed, 2019) to analyze and track gym activities accurately, providing users with real-time feedback. Modern algorithms are employed and explained through flowcharts to ensure both engineers and users can comprehend the system's complex functionality.

Combining advanced technology with a user-friendly interface, the proposed system aims to revolutionize the workout experience. It is not just a system; it is a seamless fusion of efficiency, simplicity, and creativity.

The system must follow the following instructions to function effectively as an AI gym tracker: First, OpenCV must be installed and properly configured on the system. This allows accurate prediction using the MediaPipe library and the connected webcam (Antonio, 2020; Samaan et al., 2022). Second, the system must be able to extract key points from the body joints. Third, the system should use NumPy and trigonometry to calculate joint angles to assess the accuracy of each movement. The system must also allow users to set a target repetition count and notify them once the target is achieved. Additionally, the system should alert users if a repetition is missed or if improper posture is detected. Finally, the system should use artificial intelligence (AI) to quickly and accurately count repetitions, enabling a comprehensive fitness assessment (Kumar et al., 2023) (Figure 15.2).

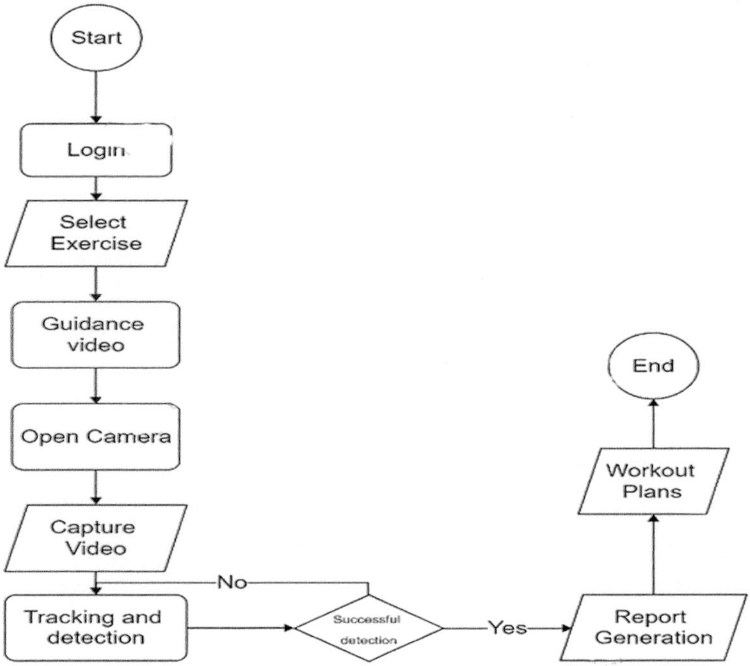

FIGURE 15.2 System flowchart.

The flowchart, which is part of the planning process, explains how the technology works for users. The system will ask users to log in via the "Start" node to authenticate the software. Once logged in successfully, they will have several options to practice. This selection of exercises forms the basis for personalized guidance delivered via instructional videos that emphasize proper form and posture. "Open Camera" is a particularly engaging step for users, as it allows the software to access the device's camera and prepare for subsequent video recording. After capturing the video, the application enters the main phase called "search and discovery" (Shitole and Devare, 2019), instantly identifying the user's body and image. As a result, users can evaluate their performance based on the generated data. Once the exercise is complete, the user can choose to continue by selecting an option. Finally, the mapping is completed, and the user's exercise is displayed by the End node. This system diagram illustrates how the body can be integrated with exercise equipment to provide users with an engaging, informative, and safe strength-training experience. These flowcharts represent the user-centered approach to enhancing the learning experience of health enthusiasts using the app. The process begins with user login, continues with exercise selection, and ends with video recording and review. Instant recognition and search technology allow users to improve their form, and the resulting reports offer insight into their progress. Finally, the exercise guide demonstrates the purpose of using technology to make workouts more effective and enjoyable. It outlines a journey that starts with a click and ends in success.

Our solution is based on an innovative and carefully designed strategy to enhance the user experience. By leveraging the power of the MediaPipe library (Samaan et al., 2022), our model aims to achieve consistency in prediction, which is critical for fitness tracking. Our system can accurately record user activity using advanced algorithms and DL. Its user-friendly interface and powerful backend architecture establish a bridge between users and the program by converting physical actions into digital data. This integration enables all movements and aspects of performance to be captured with high precision (Figure 15.3).

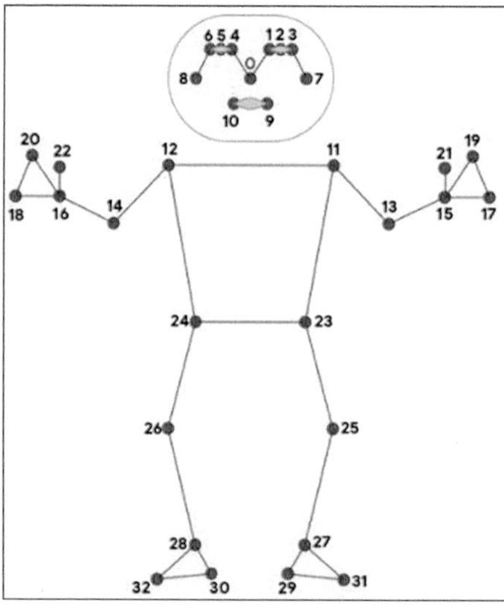

FIGURE 15.3 Pose detection image using MediaPipe library.

We integrated a CNN to improve the accuracy and reliability of the model. This advanced neural network architecture identifies subtle human movement patterns that may be missed by the human eye. By incorporating CNNs, our system can accurately detect the user's posture and movement during physical activity.

Figure 15.4 shows a flowchart of the algorithm, illustrating a simplified process for generating push notifications and tracking movements. The "Start" node represents the beginning of the algorithm and its execution sequence. The first stage, called the "Introduction video," involves the algorithm receiving a video stream, usually recorded during physical activity or exercise. This video serves as the foundation for further analysis. After the initial phase, the process is divided into two main parts—though both are labeled "Product Sales," which appears to be an error. In the "Object Detection" stage, the algorithm locates and identifies key objects or body regions within the frame. This step is particularly important for posture analysis, as it enables the algorithm to identify the user's body or specific body parts. At this point, the "Search for Items" process begins, and the program starts tracking the identified elements over time. Through this type of tracking, the system can monitor changes in body movement throughout the video. The algorithm continuously evaluates the user's posture and body position while processing the video data. If any discrepancies or issues are detected, the system triggers a "body-aware video output." This output may contain visual cues to alert users about posture issues and provide guidance on corrective actions. The behavioral monitoring process concludes when the algorithm reaches the "End" node (Nadeem, Jalal, and Kim, 2020; Pavllo et al., 2019).

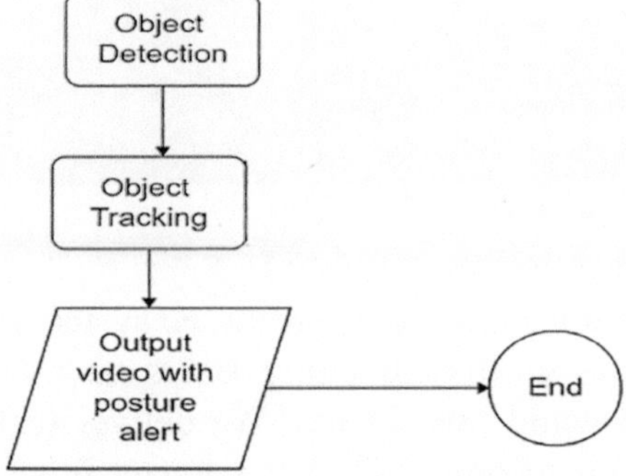

FIGURE 15.4 Algorithm flowchart.

In summary, our work explores the integration of two cutting-edge technologies—MediaPipe (Samaan et al., 2022) and CNN—to develop a model that prioritizes user safety. This system enables users to achieve their fitness goals with precision and confidence by offering real-time, accurate feedback.

15.4 RESULT AND ANALYSIS

With the help of our user-friendly interface, users can choose the exercise they want to perform after identifying the specific body area they wish to train within the proposed system. With guidance in this area, individuals can quickly and efficiently tailor their routines to meet their personal fitness objectives. Our program offers exercises based on users' input related to their selected body area, rather than requiring them to scroll through a lengthy list of activities. By selecting the corresponding YouTube video link, users can easily access visual demonstrations of each exercise. This feature improves workout planning, enhances the overall user experience, and empowers users to make more informed decisions regarding their fitness routines (Figures 15.5 and 15.6).

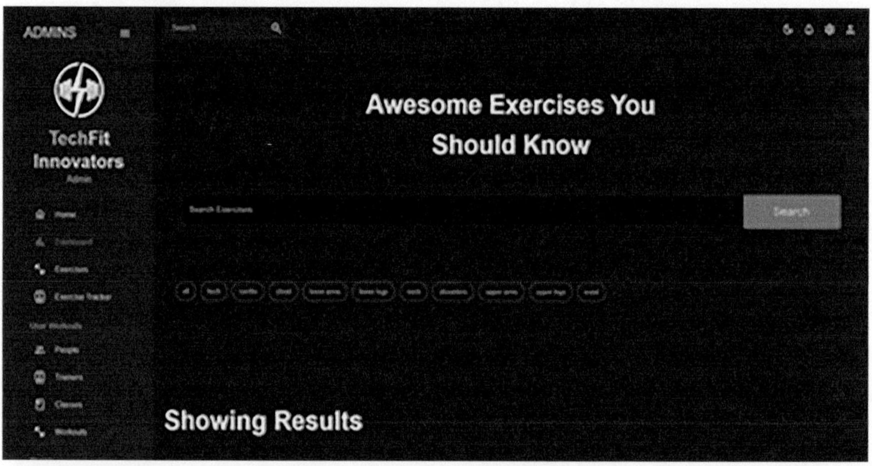

FIGURE 15.5 Exercise search UI.

Along with training advice, our web-based system for monitoring gym activity offers helpful real-time tracking. Users can monitor their progress while working out using this feature. We deliver real-time information about users' sets, repetitions, and other relevant fitness markers through tracking technology. Using this real-time feedback, users can adjust their

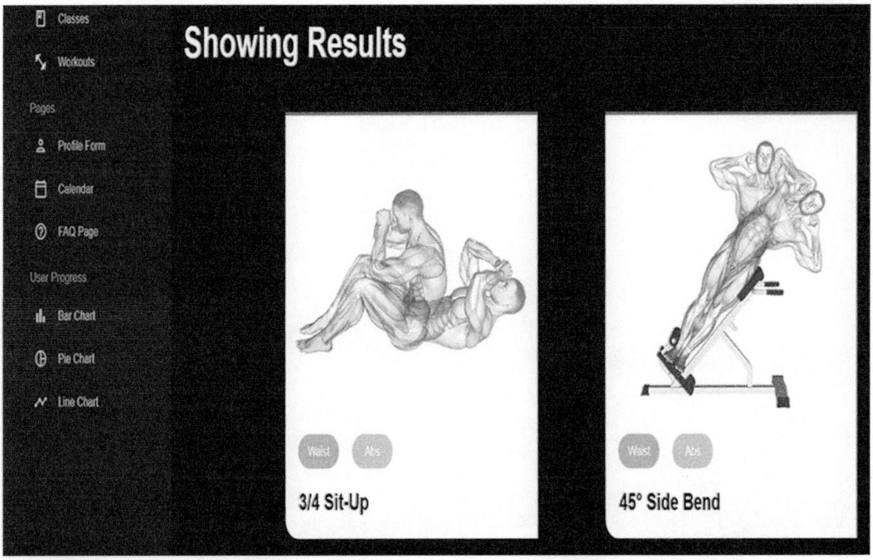

FIGURE 15.6 Exercise search results.

exercises instantly to maximize their workout and stay within their desired training zone. Real-time progress tracking enhances safety and makes training more effective by helping individuals better understand their effort levels (Figures 15.7–15.9).

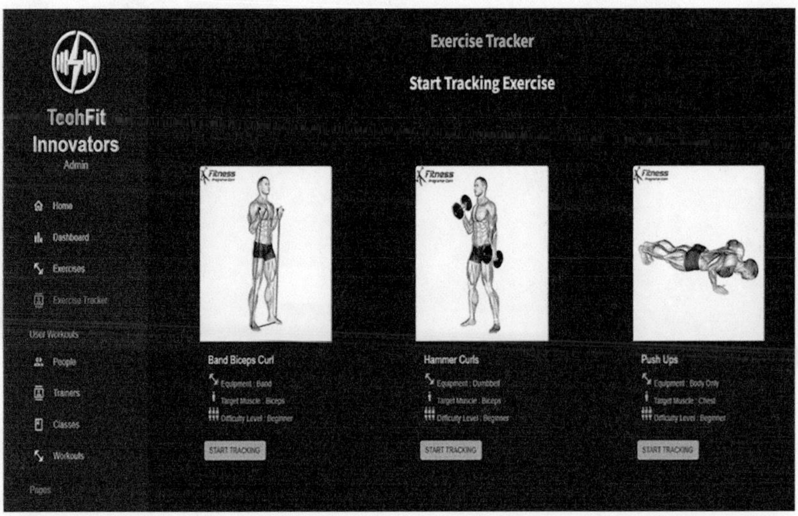

FIGURE 15.7 Exercise selection cards for monitoring.

FIGURE 15.8 Repetition counts (up).

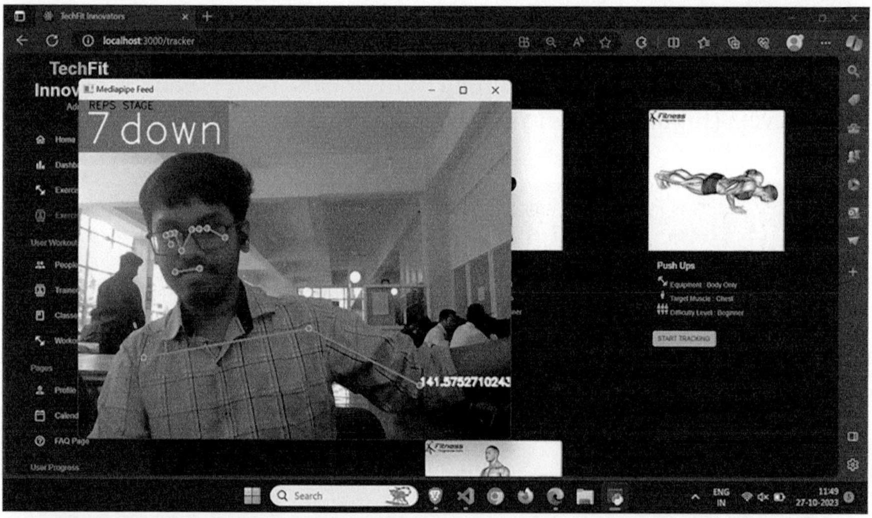

FIGURE 15.9 Repetition counts (down).

With an emphasis on progress tracking, this web application for recording gym workouts provides users with a practical tool for reaching their fitness goals and visualizing their progress. With this feature, users can log their workouts on the platform, including sets, reps, weights used, and rest periods.

The system combines rigorous analysis with ease of use, enabling users to monitor their fitness development over time. Users can create realistic goals, monitor their results, and dynamically personalize their training regimens. This instant progress tracking method promotes both accountability and motivation while fostering a sense of accomplishment, as users can view their progress in real time (Figures 15.10 and 15.11).

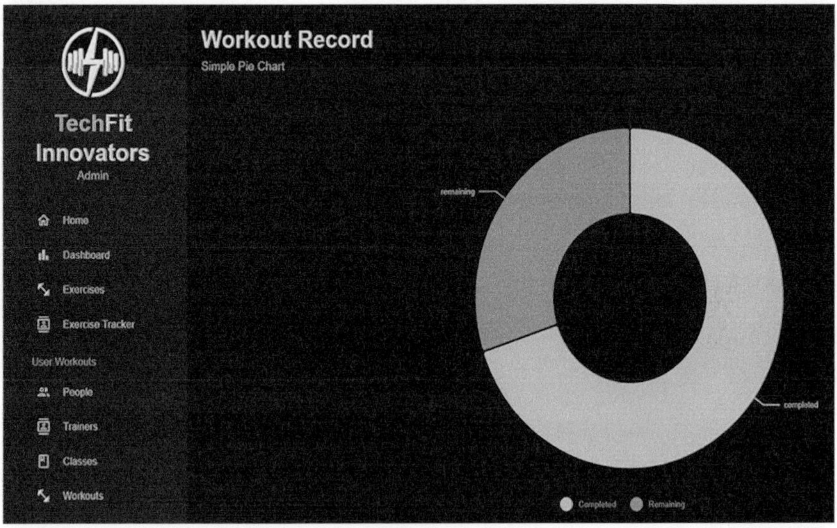

FIGURE 15.10 Workout progress tracking.

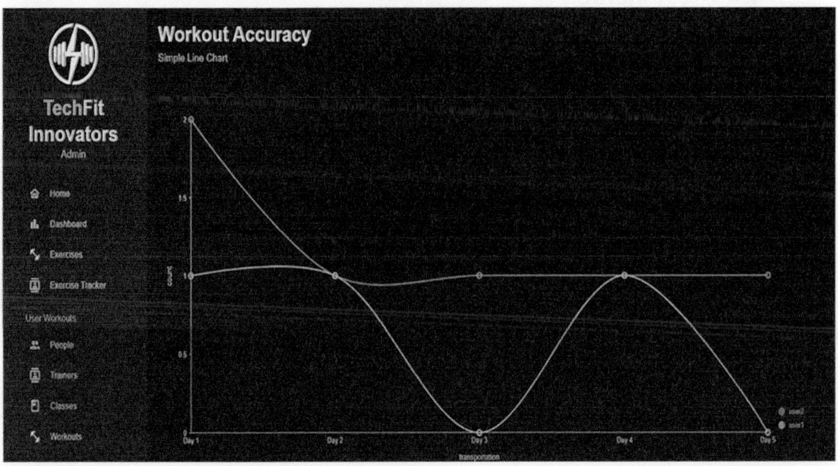

FIGURE 15.11 User data visualization.

15.5 CONCLUSION

Through improved safety and effectiveness, we believe that this innovative technique will make exercise regimens healthier and more beneficial for individuals in general good health. Education, athletics, health, fitness, and related disciplines have all benefited significantly from activity recognition technology. It opens new opportunities for enhancing user interfaces and improving communication between people and machines. Its ability to capture and analyze human body motions and postures in real time has proven valuable in enabling physical correction. Beyond simple remote workouts, it also encourages users to follow safe and efficient training routines. User data is protected, and legal compliance is ensured through the application of strong security and privacy settings.

DECLARATION

No AI tools or any other similar tools have been used in the content creation of this chapter.

KEYWORDS

- **financial identification**
- **fitness companion**
- **fitness technology**
- **personalized workouts**
- **physical correction**
- **real-time feedback**
- **social motivation**
- **user interfaces**

REFERENCES

Alam, E., Sufian, A., Dutta, P., & Leo, M. (2022). Vision-based human fall detection systems using deep learning: A review. *Computers in Biology and Medicine, 146*, 105626.

Antonio, D. L. (2020). Sign language recognition: ASL recognition with MediaPipe and recurrent neural networks. *FH Aachen University of Applied Sciences*.

Bertugl, A. I., Calderara, S., Coscia, P., Ballan, L., & Cucchiara, R. (2021). AC-VRNN: Attentive Conditional-VRNN for multi-future trajectory prediction. *Elsevier*.

Cai, Y., Ge, L., Liu, J., Cai, J., Cham, T. J., Yuan, J., & Thalmann, N. M. (2019). Exploiting spatial-temporal relationships for 3D pose estimation via graph convolutional networks. In *Proceedings of the IEEE International Conference on Computer Vision* (pp. 2272–2281).

Cao, Z., Hidalgo, G., Simon, T., Wei, S., & Sheikh, Y. (2019). OpenPose: Realtime multi-person 2D pose estimation using part affinity fields. *IEEE Transactions on Pattern Analysis and Machine Intelligence (TPAMI)*.

Chen, T., Fang, C., Shen, X., Zhu, Y., Chen, Z., & Luo, J. (2021). Anatomy-aware 3D human pose estimation with bone-based pose decomposition. In *Proceedings of the IEEE*.

Chen, X., Lin, K. Y., Liu, W., Qian, C., & Lin, L. (2019). Weakly-supervised discovery of geometry-aware representation for 3D human pose estimation. In *Proceedings of the IEEE Conference on Computer Vision and Pattern Recognition, 10*, 895–904.

Guler, R. A., & Kokkinos, I. (2020). HoloPose: Holistic 3D human reconstruction in-the-wild. In *CVPR*.

Holjevac, M., & Jakopec, T. (2020). Web application dashboards as a tool for data visualization and enrichment. *Faculty of Humanities and Social Sciences/Department of Information Sciences, Osijek*.

Javeed, A. (2019). Performance optimization techniques for ReactJS. *Springer*.

Ke, L., Pratama, K., Tai, Y. W., Tang, C. K., & Cheng, K. T. (2020). Cascaded deep monocular 3D human pose estimation with evolutionary training data. In *Proceedings of the IEEE/CVF Conference on Computer Vision and Pattern Recognition* (pp. 6173–6183).

Kocabas, M., & Athanasiou, N. (2020). Video inference for human body poses and shape estimation. In *CVPR*.

Kocabas, M., Karagoz, S., & Akbas, E. (2019). Self-supervised learning of 3D human pose using multi-view geometry. In *Proceedings of the IEEE Conference on Computer Vision and Pattern Recognition* (pp. 1077–1086).

Kokic, M., Kragic, D., & Bohg, J. (2019). Learning to estimate pose and shape of hand-held objects from RGB images. In *IROS*.

Kolotouros, N., Pavlakos, G., Michael, J. B., & Daniilidis, K. (2019). Learning to reconstruct 3D human pose and shape via model-fitting in the loop. In *ICCV*.

Kumar, R., Bajpai, A., Shina, A., & Singh, S. K. (2023). MediaPipe and CNNs for real-time ASL gesture recognition. *Elsevier*.

Li, J., Xu, C., Chen, Z., Bian, S., Yang, L., & Lu, C. (2021). Hybrik: A hybrid analytical-neural inverse kinematics solution for 3D human pose and shape estimation. In *Proceedings of the IEEE/CVF Conference on Computer Vision and Pattern Recognition*, Nashville, TN, USA (pp. 3383–3393).

Liu, S., Sehgal, N., & Ostadabbas, S. (2022). Adapted human pose: Monocular 3D human pose estimation with zero real 3D pose data. In *Proceedings of the IEEE*.

Liuhao, G., Zhou, R., Yuncheng, L., Zehao, X., Wang, Y., Cai, J., & Yuan, J. (2019). 3D hand shape and pose estimation from a single RGB image. In *Proceedings of the IEEE Conference on Computer Vision and Pattern Recognition* (pp. 10833–10842).

Ma, L., Liu, L., Theobalt, C., & Gool, V. (2022). Direct dense pose estimation. In *Proceedings of the IEEE*.

Nadeem, A., Jalal, A., & Kim, K. (2020). Human actions tracking and recognition based on body parts detection via artificial neural network. In *Proceedings of the IEEE*.

Pavllo, D., Feichtenhofer, C., Grangier, D., & Auli, M. (2019). 3D human pose estimation in video with temporal convolutions and semi-supervised training. In *Proceedings of the IEEE Conference on Computer Vision and Pattern Recognition* (pp. 7753–7762).

Samaan, G. H., Wadie, A. R., Attia, A. K., Asaad, A. M., Kamel, A. E., Slim, S. O., Abdallah, M. S., & Cho, Y. I. (2022). MediaPipe's landmarks with RNN for dynamic sign language recognition. *Electronics, 11*, 3228.

Schlett, T., Rathgeb, C., & Busch, C. (2021). Deep learning-based single image face depth data enhancement. *Elsevier*.

Shitole, A. S., & Devare, M. H. (2018). Optimization of person prediction using sensor data analysis of IoT enabled physical location monitoring. *Journal of Advanced Research in Dynamical & Control Systems, 10*(09).

Shitole, A. S., & Devare, M. H. (2019). Supervised machine learning supported time series prediction and analysis of IoT enabled physical location monitoring. *International Journal of Innovative Technology and Exploring Engineering (IJITEE), 8*, 2278–3075.

Shitole, A. S., & Devare, M. H. (2019). TPR, PPV and ROC based performance measurement and optimization of human face recognition of IoT enabled physical location monitoring. *International Journal of Recent Technology and Engineering (IJRTE), 8*(2), 2277–3878.

Verma, A., Kapoor, C., Mishra, B., & Sharma, A. (2021). Web application implementation using machine learning. In *2nd International Conference on Intelligent Engineering and Management (ICIEM)*.

Verma, A., Kapoor, C., Sharma, A., & Sharma, B. (2021). Web application implementation with machine learning. In *Proceedings of the International Conference on Innovations in Electrical Engineering and Manufacturing (ICIEM)*.

Wand, B. T., & Rosenhahn, B. (2019). RepNet: Weakly supervised training of an adversarial reprojection network for 3D human pose estimation. In *Proceedings of the IEEE Conference on Computer Vision and Pattern Recognition* (pp. 7782–7791).

Wang, J., Tan, S., Zhen, X., Xu, S., Zheng, F., He, Z., & Shao, L. (2021). Deep 3D human pose estimation. *IEEE*.

Yana, H., Gul, V., Dimitrios, T., Kalevatykh, I., Black, M. J., Laptev, I., & Schmid, C. (2019). Learning joint reconstruction of hands and manipulated objects. In *Proceedings of the IEEE Conference on Computer Vision and Pattern Recognition* (pp. 11807–11816).

CHAPTER 16

Enhancing Heart Failure Survivability Prediction: A Machine Learning Approach Using SMOTE and a Decorate Ensemble of a C4.5 Decision Tree

KALYANI KADAM,[1,*] RAHUL JOSHI,[1] JAMBI RATNA RAJA KUMAR,[2] and KISHOR RENUKADASRAO PATHAK[3]

[1]*Symbiosis Institute of Technology, Symbiosis International (Deemed University), Pune, Maharashtra, India*

[2]*Genba Sopanrao Moze College of Engineering, Balewadi, Pune, Maharashtra, India*

[3]*Vishwakarma Institute of Information Technology, Pune, Maharashtra, India*

*Corresponding author

ABSTRACT

Heart failure (HF) is a severe condition with significant mortality rates, making early survival prediction essential. In this context, machine learning (ML) plays a crucial role. This article presents an innovative ML approach using an ensemble of C4.5 decision trees (DTs) to predict survival. The C4.5 ensemble was developed within diverse ensemble creation by oppositional relabeling of artificial training examples (DECORATE), a modern ensemble method. The prediction model considers various clinical features, such as age, diabetes status, anemia status, ejection fraction, creatinine phosphokinase (CPK) levels, serum sodium, serum creatinine, high blood pressure (BP), platelet count, sex, and smoking habits, from a

Intelligent Computing and Analytics in Image Processing. Samarjeet Borah, Jambi Ratna Raja Kumar, Laxmi Bewoor, Rahul Raghvendra Joshi, and Mahendra Prabhakar Deore (Eds.)

DOI: 10.1201/9781779640772-16

dataset of 299 patients (203 survived, 96 expired) to ensure precise survival predictions. The synthetic minority oversampling technique (SMOTE) was employed to effectively balance the class imbalance in the dataset. The proposed SMOTE+DECORATE model demonstrated a classification accuracy of 86.33%, a sensitivity of 87.91%, and a specificity of 84.98%. Thus, integrating SMOTE with the DECORATE ensemble framework significantly enhances the accuracy of HF survivability predictions.

16.1 INTRODUCTION

Heart failure (HF) represents a major global health issue and is defined by the heart's inability to pump blood effectively to meet the body's demands. This condition often results from various cardiovascular diseases, such as coronary artery disease, high blood pressure (BP), and previous heart attacks (Caplan, 2023; Fayad, Robson, and Fuster, 2024). The global prevalence of HF is substantial, with millions of individuals affected, leading to considerable morbidity and mortality. HF affects approximately 6.7 million Americans aged 20 and older, with the incidence expected to rise to 8.5 million by 2030 (Bozkurt et al. 2023). Furthermore, the 5-year survival rate for individuals diagnosed with HF is approximately 26–52%, which is comparable to that of some cancers (Askoxylakis et al., 2010). These alarming statistics underscore the critical need for early detection and effective management of HF to improve patient outcomes.

Early survival prediction in HF patients is essential for several reasons. First, HF is a progressive condition that can deteriorate rapidly if not managed appropriately. Timely detection allows for early intervention, effectively slowing disease progression. This, in turn, can help reduce the frequency and duration of hospitalizations and improve the overall quality of life for patients (Abdellatif et al., 2022). With early diagnosis, patients can receive comprehensive care, including lifestyle modifications, medications, and possibly surgical interventions tailored to their specific needs. This proactive approach can prevent the condition from reaching advanced stages, where treatment options become limited and less effective. Moreover, early survival prediction using advanced tools such as machine learning (ML) models can provide precise and personalized prognostic information (Mahesh et al., 2022). ML models can analyze vast amounts of clinical data to identify patterns and risk factors associated with HF. For instance, studies might use various clinical features—such as ejection fraction, anemia status, diabetes status, creatinine phosphokinase (CPK) levels, platelet counts, high BP, serum sodium and creatinine levels, sex, and smoking habits—to predict survival (Chicco and Jurman, 2020). These models can process complex interactions between multiple variables, providing more accurate predictions than traditional statistical methods.

In HF prognosis, a significant challenge is the imbalance between patients who survive and those who do not in clinical datasets. This imbalance can lead to biased predictions and reduced model accuracy (Pes, 2020). Techniques such as data resampling can be employed to address this issue by generating synthetic instances of the minority class, thereby balancing the dataset (Barfungpa, Samantaray, and Sarma, 2024). When combined with modern ensemble frameworks, these methods can significantly improve predictive performance, offering reliable predictions that can guide clinical decision-making. The ability to predict survival accurately also aids in stratifying patients based on their risk levels. High-risk patients can be monitored more closely and provided with intensive care, while those at lower risk can be managed with less aggressive interventions. This stratification ensures the optimal allocation of healthcare resources, reducing the burden on healthcare systems and improving patient outcomes.

This study introduces an innovative ML method for predicting survival rates using a C4.5 DT ensemble. The ensemble is part of a framework known as diverse ensemble creation by oppositional relabeling of artificial training examples (DECORATE). Clinical data from 299 patients (203 surviving, 96 deceased), including age, anemia, CPK levels, diabetes status, ejection fraction, hypertension status, platelet count, serum creatinine, serum sodium, sex, and smoking habits, were analyzed for precise survival predictions. To handle data imbalance, the synthetic minority oversampling technique (SMOTE) was effectively utilized.

The contributions of the article are as follows:

- This chapter introduces an innovative machine learning approach using a DECORATE ensemble of C4.5 DTs for predicting survivability in HF patients.
- The study demonstrates the effective handling of class imbalance in the dataset using the SMOTE technique, enhancing prediction accuracy.
- Detailed performance metrics were obtained via 10-fold cross-validation.

16.2 LITERATURE REVIEW

Many articles have been proposed for predicting survival from HF data. Recently, an article used artificial intelligence (AI) to improve the prediction and understanding of HF survival outcomes (Moreno-Sánchez, 2023). Two HF survival prediction models were analyzed and evaluated using a dataset of 299 patients. The first model applies survival analysis to predict death events over time, while the second model uses a classification approach to predict mortality. This study identified the most balanced explainable prediction models by employing an optimized data workflow pipeline and various post hoc explainability techniques: the survival gradient boosting model for survival analysis and the random forest model for classification. Key clinical features such as "serum_creatinine," "ejection_fraction," and "sex" were highlighted as significant predictors in both models, enhancing interpretability and supporting potential clinical adoption.

Similarly, a support vector machine (SVM) is an AI strategy that significantly influences survivability prediction. For example, one study used a clinically validated dataset to evaluate the effectiveness of SVM and naive Bayes algorithms in predicting heart disease and patient survival (Patel et al., 2023). The naive Bayes model achieved 87% accuracy for heart disease prediction and 93% accuracy for heart survival, with an AUC of 0.93—compared to the SVM's 88% accuracy and AUC of 0.91. These findings highlight the potential of both models to improve medical decision-making by offering efficient and accurate cardiovascular disease predictions.

Like SVM, DT and logistic regression (LR) algorithms have also been optimized for predicting HF survival. DT and KNN methods achieved 10% and 3% greater accuracy, respectively, compared to previous studies (Karaoglu et al., 2021). The optimized KNN method notably assists physicians in making more effective clinical decisions.

Recently, a study addressed the urgent need for effective ECG-based AI models to predict HF risk early (Butler et al., 2023). This study validated six HF risk prediction models using multi-ethnic study of atherosclerosis (MESA) data, focusing on HF with preserved ejection fraction (HFpEF) and HF with reduced ejection fraction (HFrEF). Among these models, the ECG-AI-Cox model, which combines ECG-AI outcomes with clinical risk factors, demonstrated superior performance, with an AUC of 0.84—outperforming traditional models such as the ARIC-HF and FHS-HF models, which had AUCs of 0.76 and 0.74, respectively. ECG-AI-Cox achieved AUCs of 0.85 for HFrEF and 0.83 for HFpEF, highlighting its robust predictive capability and practical utility in clinical settings over existing approaches. Another study addressed the challenge of predicting HF mortality using a deep learning (DL) system that effectively handles missing values, high-dimensional data, and imbalances (Li et al., 2023). The system improves feature extraction and overall performance by utilizing an indicator vector and convolutional neural networks (CNNs) with multihead self-attention. The focal loss function is employed to manage data imbalances. Using the MIMIC-III dataset of 10,311 patients, the system predicts four mortality outcomes, demonstrating effectiveness through deep SHAP analysis, which identifies the top 15 predictive characteristics, enhancing medical service quality.

One study used ML to predict HF patient survival by analyzing a dataset of 299 patients (Chicco and Jurman, 2020). This study applied various classifiers to identify the most important features for survival prediction and compared the results with those of traditional biostatistics tests. Both methods identified serum creatinine and ejection fraction as the top risk factors. Models built using these two features alone performed better than those built using the full feature set, showing higher predictive accuracy. The LR model achieved 83.3% accuracy, 78% sensitivity, and 85.6% specificity. These findings suggest that focusing on serum creatinine and ejection fraction can improve HF patient survival predictions, providing a valuable tool for physicians.

Another study on the same dataset proposed a ML model to predict heart disease by incorporating diverse methods for data collection, preprocessing, and transformation (Ghosh et al., 2021). The model utilizes combined datasets (Cleveland, Long Beach VA, Switzerland, Hungarian, and Statlog) and selects features using Relief and LASSO techniques. Hybrid classifiers, such as the decision tree bagging method (DTBM), random forest bagging method (RFBM), K-nearest neighbor (KNN) bagging method (KNNBM), AdaBoost boosting method (ABBM), and gradient boosting method (GBBM), were developed by integrating traditional classifiers with bagging and boosting methods. Performance metrics were calculated, including accuracy, sensitivity, precision, F1 score, negative predictive value, false positive rate (FP Rate), and false negative (FN) rate. The model achieved the highest accuracy of 99.05% using the RFBM and Relief feature selection methods, demonstrating its effectiveness in predicting heart disease.

16.3 MATERIALS AND METHODS

The methodology relies on the SMOTE and a C4.5 DT ensemble using the DECORATE framework. Overall, the method entrusted dual-stage class balancing: first through SMOTE and then through DECORATE for classification. This dual-stage balancing enhances the robustness of the prediction model.

16.3.1 UNDERSTANDING CLASS IMBALANCE ISSUES

Class imbalance is a common challenge in ML and data mining, particularly in classification tasks where one class of interest is significantly underrepresented compared to others (Guo et al., 2008). This imbalance can skew the learning process and affect the predictive performance of models, especially when the minority class (the class of interest) contains critical or rare instances that are important for correct identification. Class imbalance occurs in a typical classification scenario when one class constitutes a much smaller proportion of the dataset than the other classes (Wang and Yao, 2012). For example, in a medical diagnosis task, the number of patients with a rare disease might be much lower than that of patients without a rare disease.

This imbalance poses several challenges, such as bias toward the majority class, where many ML algorithms tend to favor the majority class because they aim to optimize overall accuracy (Ali et al., 2019). As a result, they may misclassify or completely ignore instances from the minority class. Another challenge is that evaluation metrics are misleading, as standard performance metrics such as accuracy can be deceptive in imbalanced datasets. For instance, a model that predicts the majority class accurately but fails to detect any instance of the minority class may still achieve high accuracy. Thus, accuracy alone is not a reliable measure of model effectiveness in such cases. Oversampling and undersampling techniques can be used as preprocessing steps to address the class imbalance issue. Oversampling involves increasing the number of instances in the minority class by generating synthetic samples or replicating existing minority class instances (Ishaq et al., 2021). By balancing the class distribution, oversampling aims to provide the model with more information from the minority class. Similarly, undersampling reduces the number of instances in the majority class to match the minority class, typically by randomly removing instances from the majority class. While undersampling can speed up training and reduce bias toward the majority class, it risks losing potentially valuable information.

16.3.2 SMOTE – A POTENTIAL TECHNIQUE FOR ADDRESSING CLASS IMBALANCE ISSUES

Learning and mining from imbalanced datasets has gained significant interest in recent years. One effective way to improve the performance of standard ML techniques on these datasets is through the synthetic generation of minority samples, a technique known as SMOTE (Chawla et al., 2002). SMOTE works by creating new minority class instances based on the feature space similarities of existing minority samples, effectively increasing minority class representation.

SMOTE has proven to be a powerful tool for handling class imbalance problems. Generating synthetic examples allows classifiers to build larger decision regions that better encompass the minority class.

Here is how SMOTE works mathematically:

- **Step 1:** Randomly select a data point from the minority class. Let us call this instance x.
- **Step 2:** Determine the k nearest neighbors to x from the minority class. The value of k is a parameter you can choose.
- **Step 3:** Generate synthetic instances:
  - **Step 3.1:** Randomly select one of the k nearest neighbors. Let us call this neighbor y.
  - **Step 3.2:** Calculate the difference between the feature vectors of x and y: $diff = y - x$.
  - **Step 3.3:** Multiply *diff* by a random number between 0 and 1: $gap = rand\ diff$.
  - **Step 3.4:** Add the resulting vector to the original instance x to create a new synthetic instance:$x_{new} = x + gap$
- **Step 4:** Repeat Steps 1–3 until the desired number of synthetic instances is generated to balance the dataset.

However, medical datasets often suffer from class imbalances. For instance, the number of patients with a rare disease will be significantly lower than that of healthy individuals.

16.3.3 DECORATE ENSEMBLE FRAMEWORK

DECORATE, an ensemble framework, represents an advanced ensemble learning technique for enriching classifier diversity and improving predictive accuracy (Melville and Mooney, 2003, 2004). At its core, DECORATE enhances the traditional ensemble approach by introducing artificially generated training examples with opposite class labels to the original dataset. This process, known as oppositional relabeling, ensures that each base classifier within the ensemble learns from the original data and from instances that have been subtly altered to represent contrasting class scenarios. By doing so, DECORATE encourages ensemble members to specialize in different aspects of the data distribution, thereby reducing the risk of overfitting and enhancing the ability of the ensemble to generalize to unseen data. Mathematically, this involves augmenting the training set with modified instances, training multiple classifiers on these augmented sets, and combining their predictions through voting or weighted averaging. The result is a robust ensemble that relies on diverse viewpoints on the data, leading to more reliable and accurate predictions across various ML tasks.

The DECORATE ensemble classification procedure is as follows:

Step 1 **Generating Artificial Examples**

For each instance x_i belonging to the training set $x = \{x_1, x_2, \dots \dots, x_n\}$

If x_i belongs to class C_j, an artificial example x'_i is generated by relabeling x_i with the opposite class label C_k, where $k \neq J$.

end

Step 2 **Ensemble Training**

Step 2.1 Train a base classifier (C4.5) on the original training set.

Step 2.2 Train additional classifiers on modified training sets that include the artificially generated examples.

Training $Set_1 = \{x_1, x_2, \dots \dots, x_n, x'_1, x'_2, \dots \dots, x'_n\}$

Training$Set_1 = \{x_1, x_2, \dots \dots, x_n, x''_1, x''_2, \dots \dots, x''_n\}$

⋮

Here, $x'_n, x''_n, \dots$ are generated artificial examples for diversity.

Step 3 **Voting or Weighted Combination**

Combine predictions from all classifiers (base and ensemble members) using voting or weighted averaging to produce the final ensemble prediction.

By introducing artificially generated examples with opposite class labels, DECORATE promotes diversity among ensemble members, which can lead to improved generalization performance. Moreover, the ensemble approach tends to be more robust against overfitting and noise in the data, as the classifiers in the ensemble focus on different aspects of the data.

16.3.4 THE DATASET

The dataset utilized in this analysis comprises medical records from 299 HF patients collected between April and December 2015 (Chicco and Jurman, 2020; Heart Failure Clinical Records, UCI Machine Learning Repository. 2020). These records were sourced from two prominent medical facilities in Faisalabad, Pakistan: the Faisalabad Institute of Cardiology and Allied Hospital. This comprehensive dataset includes diverse demographic and clinical information about HF studies. The cohort comprised 105 female and 194 male patients aged 40 to 95 years. All participants were diagnosed with left ventricular systolic dysfunction, had experienced previous HF, and were categorized as class III or IV according to the New York Heart Association (NYHA) classification system. This classification is crucial for understanding the severity of their conditions, as it indicates advanced stages of HF, where patients experience significant limitations in physical activity.

The dataset encompasses 13 distinct features, offering a multifaceted view of each patient's medical profile. Among these features, several are binary, indicating the presence or absence of specific conditions or habits. These binary features include the following:

1. **Anemia:** Determined based on hematocrit levels, with a threshold set at 36%. Patients with levels below this threshold were considered anemic.
2. **High Blood Pressure (BP):** This is a critical risk factor for heart disease, and this feature indicates whether a patient has been diagnosed with hypertension.
3. **Diabetes:** Reflecting the presence of diabetes mellitus, a condition that significantly impacts cardiovascular health.
4. **Sex:** Categorized as male or female, providing essential demographic information.

5. **Smoking:** Denotes whether the patient is a smoker, an important lifestyle factor influencing heart health.

   In addition to these binary features, the dataset includes other crucial medical parameters:

6. **Age:** Patient age is essential for diagnosing and managing heart failure.
7. **Creatinine Phosphokinase (CPK):** The levels of this enzyme can indicate muscle damage, including heart muscle damage.
8. **Ejection Fraction:** This measure of the percentage of blood leaving the heart each time it contracts is vital for assessing heart function.
9. **Platelets:** The platelet count in the blood is important for understanding clotting and potential bleeding issues.
10. **Serum Creatinine:** Higher levels can indicate impaired kidney function, which is common in heart failure patients.
11. **Serum Sodium:** Levels of sodium in the blood are crucial for assessing the balance of electrolytes and overall fluid balance in the body.

These features are essential for understanding the patient's health status and risk profiles, aiding in developing targeted treatment strategies and improving prognostic assessments. Figure 16.1 presents the detailed structure and class representation of the dataset.

16.4 RESULTS AND ANALYSIS

The proposed SMOTE+DECORATE method has undergone rigorous validation through multiple cross-validation techniques to ensure its robustness and reliability. Specifically, the method was evaluated using holdout cross-validation, 10-fold cross-validation, and leave-one-subject-out (LOSO) cross-validation. These diverse validation approaches comprehensively assess the model's performance across different scenarios and data partitions. The results of these evaluations are meticulously documented, highlighting key performance metrics, including accuracy, sensitivity, and specificity. Accuracy measures the overall correctness of the model's predictions; sensitivity (or recall) assesses the model's ability

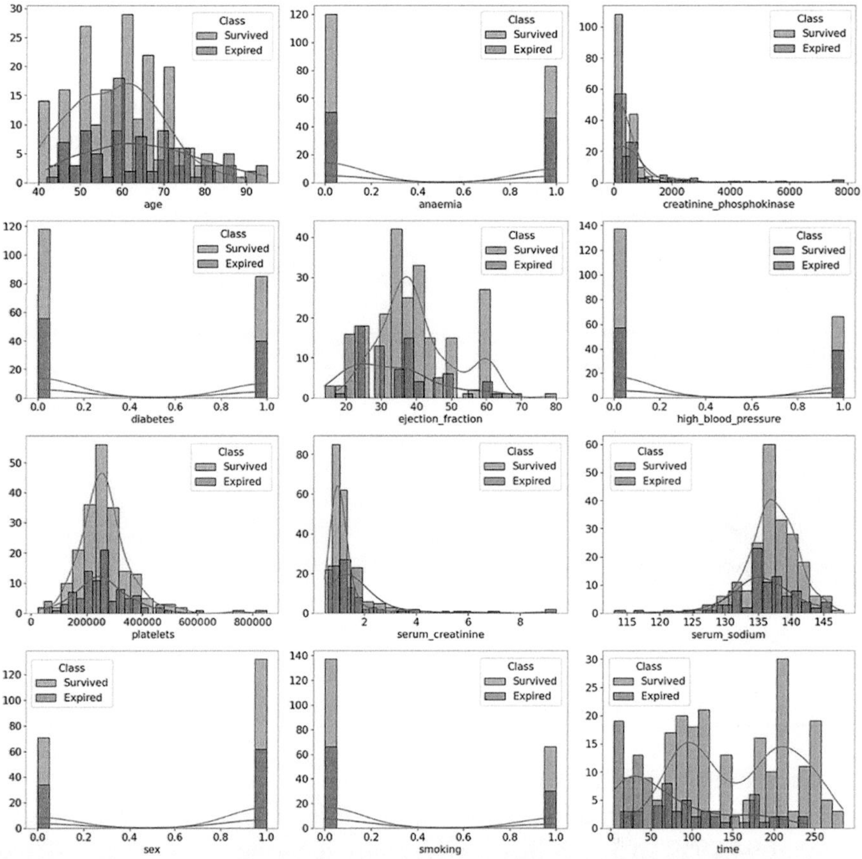

FIGURE 16.1 The structure of the heart failure dataset used with the class representation.

to identify positive instances correctly; and specificity evaluates its ability to identify negative instances correctly. Together, these metrics provide a well-rounded understanding of the model's efficacy in distinguishing between different outcomes. In addition to the standard performance metrics, the effectiveness of the proposed model was separately analyzed in terms of predicting patient survival and mortality (expired). This detailed analysis allows a nuanced understanding of the model's ability to predict these critical healthcare outcomes. By independently documenting the survivability and mortality performance, this research provides clear insights into the model's strengths and areas for potential improvement in clinical decision-making contexts.

16.4.1 HOLDOUT CROSS-VALIDATION

In holdout cross-validation, the dataset is divided into two distinct sets: a training set and a testing set. For this specific evaluation of the SMOTE+DECORATE method, 80% of the instances were allocated to the training set, while the remaining 20% were reserved for the testing set. The overall results of the classifier are presented in Table 16.1.

TABLE 16.1 Performance of the Proposed SMOTE+DECORATE Model under Holdout Cross-Validation

Performance Measures	Results
Accuracy	88.61%
Sensitivity	94.29%
Specificity	84.09%
Precision	82.50%
Negative predictive value	94.87%
False positive rate	15.91%
False discovery rate	17.50%
False negative rate	05.71%
F1 score	88.00%
Matthews correlation coefficient	77.87%

According to Table 16.1, the model achieved an accuracy of 88.61%, indicating that it correctly classified most instances. It has a high sensitivity of 94.29%, meaning that it effectively identifies positive cases, and a specificity of 84.09%, indicating good performance in recognizing negative cases. The precision is 82.50%, reflecting the accuracy of the positive predictions, while the negative predictive value is 94.87%, indicating the reliability of the negative predictions. The FP Rate is 15.91%, the false discovery rate is 17.50%, and the FN rate is 5.71%, highlighting the error rates. The F1 score is 88.00%, balancing the precision and recall, and the Matthews correlation coefficient (MCC) is 77.87%, which reflects the overall classification quality.

16.4.2 10-FOLD CROSS-VALIDATION

Table 16.2 presents the performance of the proposed SMOTE+DECORATE model evaluated using 10-fold cross-validation, which is compared to the holdout cross-validation results in Table 16.1.

TABLE 16.2 Performance of the Proposed SMOTE+DECORATE Model under 10-Fold Cross-Validation

Performance Measures	Results
Accuracy	86.33%
Sensitivity	87.91%
Specificity	84.98%
Precision	83.33%
Negative predictive value	89.16%
False positive rate	15.02%
False discovery rate	16.67%
False negative rate	12.09%
F1 score	85.56%
Matthews correlation coefficient	72.69%

The model's accuracy slightly decreases to 86.33%, indicating a small decline in overall correct classifications compared to the 88.61% achieved in the holdout validation. The sensitivity decreased to 87.91%, suggesting that the model is slightly less effective at identifying positive cases than the 94.29% in Table 16.1. The specificity remained similar at 84.98%, close to the 84.09% previously observed. The precision increases slightly to 83.33% from 82.50%, indicating a marginal improvement in the accuracy of positive predictions. The negative predictive value decreased to 89.16% from 94.87%, showing reduced reliability in negative predictions. The FP Rate is 15.02%, nearly identical to the 15.91% in Table 16.1, while the false discovery rate is 16.67%, marginally better than 17.50%. The FN rate increases to 12.09% from 5.71%, indicating more missed positive cases. The F1 score decreases to 85.56% from 88.00%, reflecting a slight decrease in the balance between precision and recall. Finally, the MCC is 72.69%, which is lower than the 77.87% in Table 16.1, suggesting a decrease in overall classification quality.

16.4.3 LEAVE-ONE-SUBJECT-OUT (LOSO) CROSS-VALIDATION

Table 16.3 shows the performance of the proposed SMOTE+DECORATE model evaluated using LOSO cross-validation and compares it with the results from holdout cross-validation (Table 16.1) and 10-fold cross-validation (Table 16.2). The accuracy in LOSO cross-validation is 84.05%, which is lower than 88.61% in holdout and 86.33% in 10-fold cross-validation, indicating a further reduction in overall correct classifications. The sensitivity is 84.86%, which is less than the 94.29% in holdout and 87.91% in 10-fold, showing the model's declining ability to identify positive cases. The specificity is 83.33%, slightly lower than 84.09% in holdout and 84.98% in 10-fold, indicating a marginal reduction in recognizing negative cases. The precision is 81.77%, compared to 82.50% in holdout and 83.33% in 10-fold, indicating a slightly lower positive predictive accuracy. The negative predictive value decreased to 86.21%, down from 94.87% in holdout and 89.16% in 10-fold, showing reduced reliability in negative predictions.

The FP Rate is 16.67%, which is greater than 15.91% in holdout and 15.02% in 10-fold, indicating more FPs. The false discovery rate is 18.23%, compared to 17.50% in holdout and 16.67% in 10-fold. The FN rate is 15.14%, which is significantly greater than 5.71% in holdout and 12.09% in 10-fold, indicating more missed positive cases. The F1 score is 83.29%, which is lower than 88.00% in holdout and 85.56% in 10-fold, reflecting a decrease in the balance between precision and recall. Finally, the MCC is 68.09%, lower than 77.87% in holdout and 72.69% in 10-fold, suggesting a further decline in overall classification quality.

TABLE 16.3 Performance of the Proposed SMOTE+DECORATE Model under Leave-One-Subject-Out (LOSO) Cross-Validation

Performance Measures	Results
Accuracy	84.05%
Sensitivity	84.86%
Specificity	83.33%
Precision	81.77%
Negative predictive value	86.21%
False positive rate	16.67%
False discovery rate	18.23%
False negative rate	15.14%
F1 score	83.29%
Matthews correlation coefficient	68.09%

16.4.4 DISCUSSION

In comparing holdout, 10-fold cross-validation, and LOSO cross-validation for evaluating the proposed SMOTE+DECORATE model, 10-fold cross-validation emerges as the most robust choice due to several key advantages. First, it strikes an optimal balance between computational efficiency and reliable performance estimation, which is particularly advantageous when working with datasets of moderate size. Unlike holdout validation, which relies on a single random dataset split into training and testing sets, 10-fold cross-validation mitigates the risk of high variance by averaging performance metrics over ten iterations. Each iteration involves training the model on nine-tenths of the data and testing it on the remaining one-tenth, ensuring that the model's performance is evaluated across multiple diverse subsets of the data. On the other hand, LOSO cross-validation exhaustively tests the model by leaving out one instance at a time for testing, making it thorough but computationally intensive, especially with larger datasets. This approach can also be more susceptible to variability when sample sizes are limited, potentially affecting the stability of performance metrics. In contrast, 10-fold cross-validation provides a more stable estimation of crucial model evaluation metrics such as accuracy, sensitivity, specificity, and F1 score. These metrics are essential for assessing the model's ability to generalize effectively across diverse datasets and real-world applications, ensuring that the findings are robust and dependable.

Therefore, considering its ability to mimic real-world application scenarios, computational feasibility, and reliability in performance estimation, 10-fold cross-validation is the preferred method for evaluating the SMOTE+DECORATE model. It enhances confidence in the model's predictive capabilities and supports informed decision-making in clinical settings, where accurate predictions of HF survivability are crucial for improving patient outcomes and healthcare management strategies.

16.5 SURVIVAL AND MORTALITY ANALYSIS

The proposed SMOTE+DECORATE model exhibited acceptable performance in predicting survival and mortality under 10-fold cross-validation, as detailed in Table 16.4. For the expired category, the model achieved a true positive rate (TP rate) of 0.833, indicating that it accurately identified 83.3% of the patients who did not survive, reflecting high sensitivity. The FP rate was low at 0.108, indicating that only 10.8% of the patients who survived were incorrectly predicted to have died. The precision, which measures the accuracy of the positive predictions, is 0.879, showing that 87.9% of the patients who were predicted to expire did not survive. The recall, equivalent to the TP rate, reiterates the model's ability to identify expired cases effectively. The F-measure, a harmonic mean of precision and recall, stands at 0.856, demonstrating balanced performance between precision and recall. The MCC, which considers true and FPs and negatives, is 0.727, indicating a strong positive correlation between the observed and predicted outcomes.

TABLE 16.4 Survivability and Mortality Prediction under 10-Fold Cross-Validation

Status	TP Rate	FP Rate	Precision	Recall	F-Measure	MCC	ROC Area	PRC Area
Expired	0.833	0.108	0.879	0.833	0.856	0.727	0.918	0.869
Survived	0.892	0.167	0.85	0.892	0.87	0.727	0.918	0.925

For the survived category, the model also showed commendable results. The TP rate was 0.892, indicating that the model correctly identified 89.2% of the surviving patients. The FP rate was 0.167, indicating that 16.7% of the patients who died were incorrectly predicted to survive. The precision was 0.850, suggesting that 85% of the patients predicted to survive actually survived. The recall is consistent with the TP rate of 0.892, reflecting the model's ability to identify surviving patients correctly. The F-measure is 0.870, indicating a well-balanced trade-off between precision and recall. The MCC for the survived category was 0.727, the same as that for the expired category, underscoring the model's overall strong predictive capability across both categories.

The impressive results of the SMOTE+DECORATE model can be attributed to its ability to effectively handle class imbalance through the SMOTE, which ensures that the model has sufficient examples from both the minority and majority classes, thereby improving its learning process. Additionally, the DECORATE ensemble technique enhances the model's robustness and generalizability by training diverse classifiers and combining their predictions. This combination leads to a more accurate and reliable model capable of providing high precision and recall for both survivability and mortality predictions. The model achieves superior performance by integrating these advanced techniques, making it a valuable tool for predicting patient outcomes in clinical settings.

To further explore survivability and mortality prediction, the receiver operating characteristic (ROC) curve for the prediction was plotted. Figure 16.2 presents the ROC curve for the proposed SMOTE+DECORATE model, depicting its performance in predicting whether patients would expire or survive under 10-fold cross-validation. The area under the curve (AUC) for both the expired and survived categories was 0.918, which is remarkably high and indicates that the model has excellent discriminatory ability. The AUC is a crucial metric that measures the overall ability of the model to distinguish between the classes (expired and survived) across all possible classification thresholds. An AUC of 0.918 means that there is a 91.8% chance that the model will correctly differentiate between a randomly chosen patient who died and a randomly chosen patient who survived. Here, the ROC curve plots the TP rate (sensitivity) against the FP rate (1 − specificity) at various threshold settings. The curve's proximity to the top-left corner of the graph indicates a high sensitivity and a low FP rate, both of which are desirable attributes in a predictive model. In the case of the expired category, the ROC curve demonstrated that the model effectively identified the majority of patients who did not survive while minimizing the number of patients who were incorrectly predicted to have expired. Similarly, for the survived category, the ROC curve showed that the model accurately identified patients who survived, with a low rate of FPs, whereas patients who died were incorrectly predicted to have survived. A high AUC reflects the model's robustness in handling the dataset's complexity and variability. This robustness is further enhanced by the 10-fold cross-validation technique, which ensures that the model's performance is consistent and not overly dependent on any specific subset of the data. Each fold of the cross-validation process contributes to the creation of a comprehensive and reliable model by training on 90% of the data and validating on the remaining 10%, thereby providing a rigorous evaluation of the model's predictive power.

Overall, Figure 16.2 reveals the effectiveness of the SMOTE+DECORATE model in achieving high accuracy and reliability in predicting patient outcomes, with a balanced trade-off between sensitivity and specificity. The impressive AUC of 0.918 for both categories underscores the model's potential as a valuable tool for clinical decision-making, aiding healthcare professionals in identifying patients at risk of mortality with a high degree of confidence. This capability is particularly important in

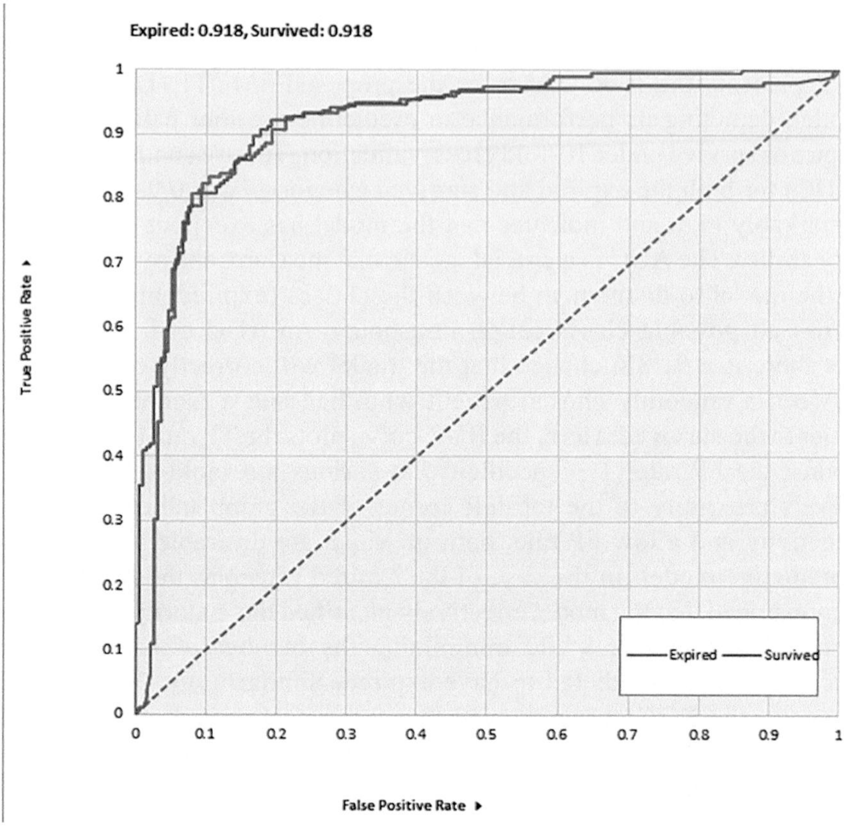

FIGURE 16.2 The receiver operating characteristic (ROC) curves of patients who were predicted to have died or survived under 10-fold cross-validation.

the context of HF, where timely and accurate predictions can significantly impact patient management and treatment strategies.

16.6 CONCLUSION

Heart failure (HF) is a critical health challenge with high mortality rates, highlighting the importance of early survival prediction. This article introduced an innovative ML approach using an ensemble of C4.5 DTs developed within the DECORATE framework. By leveraging clinical data such as age, anemia status, and various biomarkers from a dataset of 299 patients, including 203 survivors and 96 nonsurvivors, the model aimed to predict survival accurately. The SMOTE technique was employed to enhance the effectiveness of model training in addressing the inherent class imbalance in the dataset. The SMOTE+DECORATE ensemble achieved promising results with an accuracy of 86.33%, sensitivity of 87.91%, and specificity of 84.98% under 10-fold cross-validation. In addition to 10-fold cross-validation, holdout cross-validation and LOSO cross-validation techniques were used to validate the proposed model.

DECLARATION

No AI tools or any other similar tools have been used in the content creation of this chapter.

KEYWORDS

- **biomarkers**
- **class imbalance**
- **cross-validation**
- **DECORATE**
- **heart failure**
- **mortality rates**
- **SMOTE**
- **survivability prediction**

REFERENCES

Abdellatif, A., Abdellatef, H., Kanesan, J., Chow, C. O., Chuah, J. H., & Gheni, H. M. (2022). Improving the heart disease detection and patients' survival using supervised infinite feature selection and improved weighted random forest. *IEEE Access, 10*, 67363–67372. https://doi.org/10.1109/ACCESS.2022.3185129 (accessed on 11th July 2025).

Ali, H., Salleh, M. N. M., Saedudin, R., Hussain, K., & Mushtaq, M. F. (2019). Imbalance class problems in data mining: A review. *Indonesian Journal of Electrical Engineering and Computer Science, 14*(3), 1560–1571.

Askoxylakis, V., et al. (2010). Long-term survival of cancer patients compared to heart failure and stroke: A systematic review. *BMC Cancer, 10*(1), 105. https://doi.org/10.1186/1471-2407-10-105 (accessed on 11th July 2025).

Barfungpa, S. P., Samantaray, L., & Sarma, H. K. D. (2024). SMOTE-based adaptive coati kepler optimized hybrid deep network for predicting the survival of heart failure patients. *Multimedia Tools and Applications*. https://doi.org/10.1007/s11042-023-18061-3 (accessed on 11th July 2025).

Bozkurt, B., et al. (2023). Heart failure epidemiology and outcomes statistics: A report of the Heart Failure Society of America. *Journal of Cardiac Failure, 29*(10), 1412–1451. https://doi.org/10.1016/j.cardfail.2023.07.006 (accessed on 11th July 2025).

Butler, L., et al. (2023). A generalizable electrocardiogram-based artificial intelligence model for 10-year heart failure risk prediction. *Cardiovascular Digital Health Journal, 4*(6), 183–190. https://doi.org/10.1016/j.cvdhj.2023.11.003 (accessed on 11th July 2025).

Caplan, R. M. (2023). Heart disease and hypertension. In *Long Life Strategy* (pp. 109–124). Springer Nature Switzerland. https://doi.org/10.1007/978-3-031-44518-7_10 (accessed on 11th July 2025).

Chawla, N. V., Bowyer, K. W., Hall, L. O., & Kegelmeyer, W. P. (2002). SMOTE: Synthetic minority oversampling technique. *Journal of Artificial Intelligence Research, 16*, 321–357.

Chicco, D., & Jurman, G. (2020). Machine learning can predict survival of patients with heart failure from serum creatinine and ejection fraction alone. *BMC Medical Informatics and Decision Making, 20*(1), 16. https://doi.org/10.1186/s12911-020-1023-5 (accessed on 11th July 2025).

Fayad, Z. A., Robson, P. M., & Fuster, V. (2024). Rethinking heart attack prevention. *Journal of the American College of Cardiology, 83*(22), 2145–2147. https://doi.org/10.1016/j.jacc.2024.04.007 (accessed on 11th July 2025).

Ghosh, P., et al. (2021). Efficient prediction of cardiovascular disease using machine learning algorithms with Relief and LASSO feature selection techniques. *IEEE Access, 9*, 19304–19326. https://doi.org/10.1109/ACCESS.2021.3053759 (accessed on 11th July 2025).

Guo, X., Yin, Y., Dong, C., Yang, G., & Zhou, G. (2008). On the class imbalance problem. In *2008 Fourth International Conference on Natural Computation* (pp. 192–201). IEEE. https://doi.org/10.1109/ICNC.2008.871 (accessed on 11th July 2025).

Heart Failure Clinical Records, UCI Machine Learning Repository. (2020). https://doi.org/10.24432/C5Z89R (accessed on 11th July 2025).

Ishaq, A., et al. (2021). Improving the prediction of heart failure patients' survival using SMOTE and effective data mining techniques. *IEEE Access, 9*, 39707–39716. https://doi.org/10.1109/ACCESS.2021.3064084 (accessed on 11th July 2025).

Karaoglu, A. N., Caglar, H., Degirmenci, A., & Karal, O. (2021). Performance improvement with decision tree in predicting heart failure. In *2021 6th International Conference on Computer Science and Engineering (UBMK)* (pp. 781–784). IEEE. https://doi.org/10.1109/UBMK52708.2021.9558939 (accessed on 11th July 2025).

Li, D., Fu, J., Zhao, J., Qin, J., & Zhang, L. (2023). A deep learning system for heart failure mortality prediction. *PLoS One, 18*(2), e0276835. https://doi.org/10.1371/journal.pone.0276835 (accessed on 11th July 2025).

Mahesh, T. R., et al. (2022). AdaBoost ensemble methods using K-fold cross validation for survivability with the early detection of heart disease. *Computational Intelligence and Neuroscience, 2022*, 1–11. https://doi.org/10.1155/2022/9005278 (accessed on 11th July 2025).

Melville, P., & Mooney, R. J. (2003). Constructing diverse classifier ensembles using artificial training examples. In *Eighteenth International Joint Conference on Artificial Intelligence* (pp. 505–510).

Melville, P., & Mooney, R. J. (2004). Creating diversity in ensembles using artificial data. *Information Fusion: Special Issue on Diversity in Multiclassifier Systems*.

Moreno-Sánchez, P. A. (2023). Improvement of a prediction model for heart failure survival through explainable artificial intelligence. *Frontiers in Cardiovascular Medicine, 10*. https://doi.org/10.3389/fcvm.2023.1219586 (accessed on 11th July 2025).

Patel, T. S., Patel, D. P., Sanyal, M., & Shrivastav, P. S. (2023). Prediction of heart disease and survivability using support vector machine and Naive Bayes algorithm. *bioRxiv*, 2006–2023.

Pes, B. (2020). Learning from high-dimensional biomedical datasets: The issue of class imbalance. *IEEE Access, 8*, 13527–13540. https://doi.org/10.1109/ACCESS.2020.2966296 (accessed on 11th July 2025).

Wang, S., & Yao, X. (2012). Multiclass imbalance problems: Analysis and potential solutions. *IEEE Transactions on Systems, Man, and Cybernetics, Part B: Cybernetics, 42*(4), 1119–1130. https://doi.org/10.1109/TSMCB.2012.2187280 (accessed on 11th July 2025).

CHAPTER 17

Utilizing Adaptive Feature Scaling and Dynamic Contextual Graph Networks to Improve Captioning of Images Through the Recognition of Unseen Relationships

YUGANDHARA A. THAKARE,[1,*] K. H. WALSE,[2] and MOHAMMAD ATIQUE[1]

[1]*P.G. Department of Computer Science, Sant Gadge Baba Amravati University, Amravati, Maharashtra, India*

[2]*Sant Bhagwan Baba Kala Mahavidyalaya, Sindkhed Raja, Maharashtra, India*

*Corresponding author

ABSTRACT

Automated systems for image captioning are increasingly important in fields like assistive technology, content management, and robotics. However, current systems often fall short when it comes to accurately capturing unseen relationships and the deeper contextual meaning in images, which limits their practical use. These systems typically depend on convolutional neural networks (CNNs) and basic attention mechanisms, which result in captions that lack semantic depth and lead to subpar performance in recognizing both objects and their relationships. This paper presents a cutting-edge approach to image captioning that combines adaptive feature scaling for enhanced object detection, dynamic contextual graph networks for improved comprehension of contextual information, meta-learning for

Intelligent Computing and Analytics in Image Processing. Samarjeet Borah, Jambi Ratna Raja Kumar, Laxmi Bewoor, Rahul Raghvendra Joshi, and Mahendra Prabhakar Deore (Eds.)

DOI: 10.1201/9781779640772-17

zero-shot detection to identify previously undiscovered relationships, and a multimodal attention mechanism to produce more detailed and precise captions. This framework demonstrates superior performance over existing models, particularly in the classification of objects and their relationships, making it highly applicable for automated social media content generation and image analysis. During testing, the model showed a 4.9% boost in object classification precision and a 4.5% increase in accuracy, along with a remarkable improvement in relationship recognition, achieving an 8.3% increase in precision and an 8.5% improvement in accuracy.

17.1 INTRODUCTION

As the volume of visual data continues to surge, the capacity to infer, define, and understand images in a natural way has become increasingly significant. This demand spans multiple domains, such as providing assistance to visually impaired individuals, boosting the functionality of autonomous robots, enhancing digital archiving, and improving social media analysis. While there have been significant strides in computer vision, a persistent challenge lies in the ability to grasp and convey complex relationships and contextual nuances in images, beyond simply identifying objects (Liu et al., 2022; Song et al., 2023; Ye et al. 2022). Many existing approaches struggle when tasked with recognizing new, unseen relationships or capturing the deeper context within an image set. This shortcoming limits the effectiveness and reliability of these systems, particularly in dynamic, real-world scenarios (Cheng et al., 2022; Ma, Zhao, and Shi, 2021; Sumbul, Nayak, and Demir, 2021).

In view of the current limitations and their considerable effect on everyday applications, this paper introduces a novel image captioning structure aimed at addressing the deficiencies of existing methods and thereby enhancing the effectiveness of cutting-edge image captioning systems.

Additionally, the demand for enhanced image captioning spans various sectors like education, where automated systems could help create more detailed and contextually rich descriptions of visual content for learning purposes. In the healthcare industry, particularly in radiology and pathology, an advanced system that can accurately interpret complex medical images and highlight subtle relationships between features could lead to better diagnostic outcomes and improved efficiency. Similarly, in the legal domain, refined image captioning could assist in analyzing crime scene photographs or evidence, offering detailed, contextual interpretations to support investigations or legal documentation.

Additionally, the advancement of AR and VR technologies has created a pressing need for image captioning systems capable of generating captions in real time, delivering immersive and interactive user experiences. As these technologies continue to evolve, the ability to describe and interpret not just static images but also dynamic visual environments will become essential. The integration of sophisticated image captioning models within AR/VR could transform user experiences, particularly in areas such as gaming, training simulations, and remote collaboration, making them more intuitive and engaging.

The ultimate aim of this research is to narrow the gap between human-like image comprehension and machine-generated captions. By incorporating advancements in feature extraction, contextual understanding, and relationship detection, the proposed framework advances the state of current image captioning techniques and lays the foundation for future innovations in AI-driven image interpretation. Designed to be adaptable across multiple domains, this framework ensures its significance and application to both current and emerging technologies.

17.2 REVIEW OF LITERATURE

The integration of computer vision and NLP to generate text descriptions that are easily understandable by humans has emerged as a crucial research focus. This integration seeks to improve machine–user interactions by delivering insightful and contextually appropriate information. The objective is to create systems capable of interpreting and describing visual data in a way that mirrors human understanding (Kim et al., 2022; Liu et al., 2022; Zha et al., 2022). While progress has been made, existing techniques each have their own strengths and weaknesses. Earlier image captioning methods relied on predefined templates, leading to generic captions that missed important contextual details and the relationships between objects (Wang et al., 2023; Wu et al., 2021). Rule-based models introduced conditional rules, but they were heavily dependent on manually created rules, which made them less effective when handling complex scenes (Haque, Ghani, and Saeed, 2021; Huang and Wang, and Li, 2021; Wang et al., 2022).

Encoder–decoder models, using LSTM or GRU networks, were effective to some extent but faced challenges with lengthy sentences and nuanced context (Nguyen et al., 2022; Wang et al., 2023; Zhang et al. 2023). Attention-based models improved by allowing selective focus on image regions but still had difficulties with complex visual relationships (Aafaq et al., 2022; Im and Chan, 2023; Yang et al., 2022).

Recent research has integrated transformer models, inspired by their success in natural language processing (NLP), into image captioning (Bae et al., 2022; Hafeth, Kollias, and Ghafoor, 2023; Nguyen and Fernando, 2022). While promising for contextual understanding, they require significant computational resources. Cutting-edge models incorporate multiple data types but depend on well-structured supplementary data.

Existing image captioning models have limitations in detecting unseen relationships, comprehending complex scenes (Liu, Zhao, and Shi, 2022; Zhuang et al., 2022), and providing meaningful context (Hinz, Heinrich, and Wermter, 2022; Wang, Wan, and Chan, 2022). To address these issues, the next section discusses the development of an effective model for dynamic contextual graph networks and adaptive feature scaling for image captioning that involves the detection of previously undiscovered relationships.

Recent progress in image captioning has benefited significantly from deep learning (DL) models, particularly through feature extraction using convolutional neural networks (CNNs) combined with long short-term memory (LSTM) networks for generating sequences. Article (Hartatik, Al Fatta, and Fajar, 2019) introduced a system utilizing these two architectures to automatically generate image captions, highlighting the strength of CNNs in capturing spatial features and LSTMs in producing fluent and meaningful textual sequences. Their results demonstrated that combining these neural networks significantly enhances caption quality in various tasks.

Study work (Sargar and Kinger, 2021) provided a comprehensive overview of different image captioning techniques and evaluation metrics, examining how diverse methods influence performance across different datasets. They discussed the evolution from traditional methods to more sophisticated DL models, emphasizing the need for a balance between accuracy and computational efficiency, especially with increasingly large and complex datasets. Their analysis also focused on key assessment metrics like BLEU, METEOR, and CIDEr, which are critical for effectively measuring the performance of captioning systems.

In a related study by Chu et al. (2020) implemented a framework combining ResNet50 for extracting detailed visual features with LSTMs equipped with a soft attention mechanism. This attention mechanism enabled the model to concentrate on particular areas of the image, resulting in captions that are more precise and contextually appropriate. Their approach demonstrated how integrating advanced CNN architectures like ResNet with sequence modeling techniques can significantly improve captioning performance, especially in complex scenarios.

17.3 PROPOSED MODEL

A review of existing image captioning models highlights their struggles with real-time application challenges and scalability concerns, particularly when dealing with complex image and video sequences. To address these challenges, this segment introduces a model that integrates adaptive feature scaling with dynamic contextual graph networks, aimed at improving the efficiency of image captioning while enabling the detection of novel relationships.

As shown in Figure 17.1, the proposed model leverages adaptive feature scaling to enhance object recognition accuracy, followed by dynamic contextual graph networks to better capture contextual relationships. Moreover, meta-learning is utilized for zero-shot detection of previously unseen relationships, and a multimodal attention mechanism is employed to optimize the captioning process.

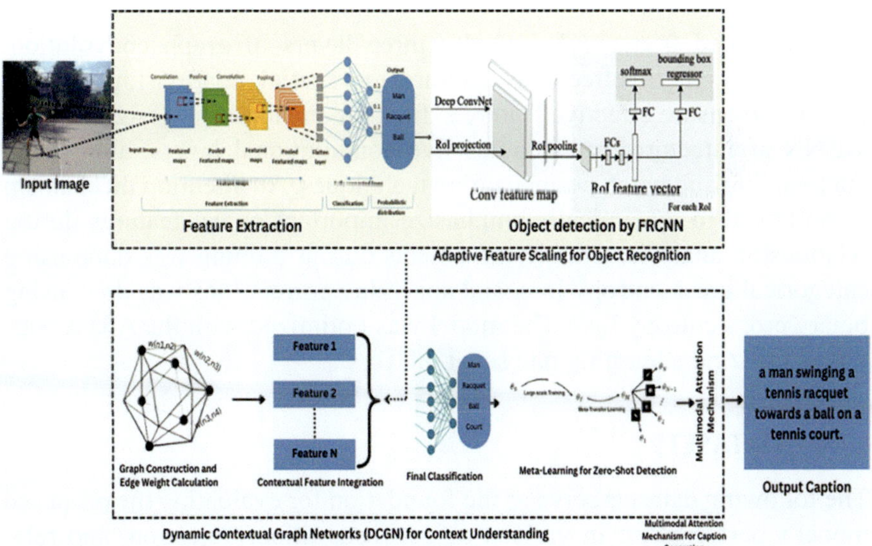

FIGURE 17.1 The proposed approach.

- **Step 1:** The image captioning process involves several key steps: First, high-level features are extracted using a pre-trained CNN model to capture essential visual information.
- **Step 2:** Then, object detection through Faster R-CNN identifies objects and their spatial relationships.
- **Step 3:** Adaptive feature scaling emphasizes crucial information by reducing the dimensionality of object characteristics.
- **Step 4:** Dynamic contextual graph networks represent relationships between objects, enhancing scene understanding.
- **Step 5:** Meta-learning enables the model to detect unseen relationships by generalizing across contexts.
- **Step 6:** A multimodal attention mechanism helps the model focus on important regions and relationships for accurate caption generation.
- **Step 7:** Finally, the model decodes these integrated features to produce a coherent, contextually relevant caption.

17.3.1 MODEL CONFIGURATION

The suggested framework includes three layers of graph convolution, which allows it to effectively capture the spatial relationships among objects. To ensure effective feature extraction, the model is designed using a CNN architecture that includes five convolutional layers, along with max pooling and batch normalization techniques. An attention mechanism is employed to dynamically emphasize important object features during relationship analysis. Classifying objects during training was done using categorical cross-entropy loss, and analyzing correlations was done using binary cross-entropy loss. The model was optimized with the Adam optimizer, utilizing a learning rate set at 0.001.

17.4 DATASETS

The following datasets serve as the foundation for evaluating the proposed model's performance in various object recognition, detection, and relationship analysis tasks.

17.4.1 COCO DATASET (COMMON OBJECTS IN CONTEXT)

A common benchmark for object detection and image segmentation, it consists of over 200,000 images, each labeled with object bounding boxes and class labels. For our experiments, we randomly chose 10,000 images to represent a wide variety of object categories.

17.4.2 PASCAL VOC DATASET

A well-known dataset for object recognition and detection tasks; it includes 20 object categories and roughly 17,000 labeled images. For our experiment, we used a subset of 5000 images from the Pascal VOC dataset.

17.4.3 IMAGENET

A large dataset containing millions of images across various categories. For our experiment, we selected a subset of 2000 images, emphasizing a range of objects and scenes.

17.4.4 VISUAL RELATIONSHIP DATASET

Created to evaluate the model's relationship analysis capabilities. This custom dataset comprises 1000 images with annotated object pairs and corresponding relationship labels. It includes relationships like "on," "under," "beside," and "inside" between objects.

17.5 ANALYSIS OF RESULTS AND COMPARATIVE EVALUATION

The proposed model represents an advanced DL architecture specifically created for object classification along with relationship analysis within intricate visual scenes. By integrating graph convolutional layers with a CNN backbone, it effectively captures spatial relationships and situational context for accurate item identification. Furthermore, a dynamic focus mechanism, incorporated to improve the model's responsiveness to real-world situations, enhances its capability to highlight relevant features during analysis. Its graph-based methodology and flexible training framework exhibit significant potential in both object classification and the examination of relationships that were previously unknown, making it a valuable tool in computer vision applications. Comprehensive experiments are presented. In this part, we assess the model's performance across various datasets, configurations, and assessment metrics. The findings highlight the model's effectiveness across a range of circumstances.

Equations (1) and (2) were employed to evaluate the levels of precision (*P*) and accuracy (*A*):

$$Precision = \frac{TP}{TP + FP} \tag{1}$$

$$Accuracy = \frac{TP + TN}{TP + TN + FP + FN} \tag{2}$$

Figure 17.2 illustrates the precision levels derived from these assessments.

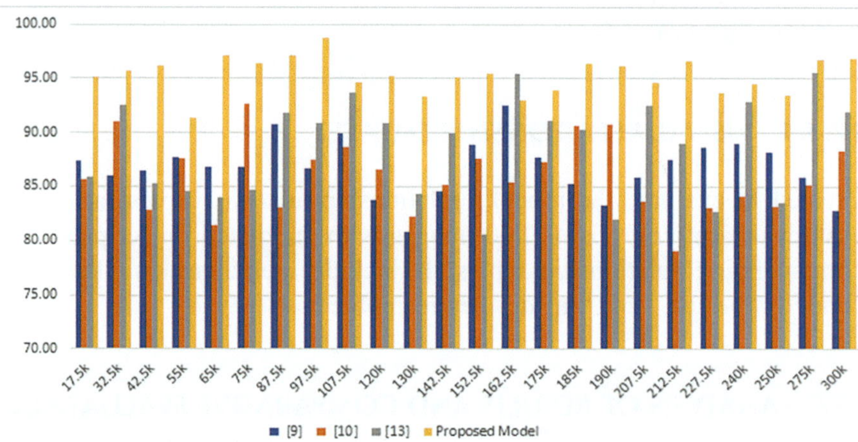

FIGURE 17.2 Precision in categorizing objects into meaningful texts.

The prime aim of this research is to analyze the accuracy of various models in categorizing objects into semantic groups using datasets of different sizes. Among the models analyzed, including Zha et al. (2022), Huang et al. (2022), Zhang et al. (2022), and the proposed model, our model consistently exhibits superior object classification precision.

For instance, when tested on a dataset with 17.5k samples, the proposed model achieves a remarkable precision rate of 95.06%, surpassing Zha et al. (2022) by 7.62%, Huang et al. (2022) by 9.48%, and Zhang et al. (2022) by 9.14%. This substantial performance gap highlights our model's exceptional accuracy in identifying and classifying objects.

As the dataset size rises, the proposed model continues to maintain its advantage. For example, utilizing a dataset comprising 97,500 samples, it accomplishes an impressive precision of 98.76%, outperforming its nearest competitor (Zhang et al., 2022), by 7.86%. This further demonstrates the robustness of our model in object classification, even when handling larger and more complex datasets.

The impressive performance of the proposed model is primarily due to its innovative elements. The scaling of adaptive features mechanism allows for dynamic adjustment of feature scales, greatly improving object recognition accuracy. By integrating dynamic contextual graph networks, the model gains a deeper understanding of context, enabling it to effectively capture intricate relationships between objects and their surroundings. Moreover, the inclusion of meta-learning for zero-shot detection empowers the model to identify previously unobserved relationships, further enhancing its accuracy over models that lack this feature.

In summary, the proposed model consistently outperforms other models such as Zha et al. (2022), Huang et al. (2022), and Zhang et al. (2022) in object classification precision across datasets of varying sizes. This benefit results from the combination of meta-learning for detection without prior examples, dynamic contextual graph networks, and adaptive feature scaling, all of which contribute to its superior accuracy in recognizing and classifying objects. The accuracy comparison of these models is shown in Figure 17.3.

Accurate classification of objects, especially in tasks like image captioning, is crucial. We are evaluating the accuracy of different models, including Zha et al. (2022), Huang et al. (2022), Zhang et al. (2022), and the newly proposed model, across datasets of various sizes.

Our model consistently surpasses other models in the accuracy of object classification, regardless of the dataset size. For example, on a dataset containing 17.5k samples, the proposed model achieves an accuracy of 91.96%, outperforming Zha et al. (2022) by 2.57%, Huang et al. (2022) by 7.83%, and Zhang et al. (2022) by 7.14%. This showcases the system's

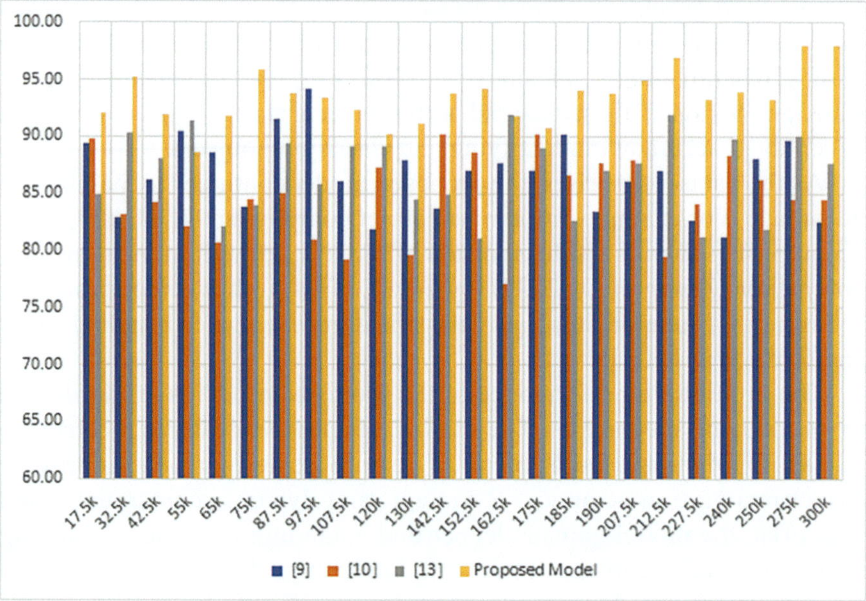

FIGURE 17.3 Accuracy in categorizing objects into semantic texts.

ability to reliably recognize objects in images, even when working with smaller datasets.

As dataset sizes increase, the proposed model maintains impressive accuracy, reaching 98.02% on a dataset containing 300,000 samples, outperforming Zhang et al. (2022) by 7.04%. This underscores the model's ability to effectively manage large and complex datasets. These features improve the model's effectiveness in object recognition, contextual understanding, and detecting unseen relationships.

In summary, the proposed model typically surpasses Zha et al. (2022), Huang et al. (2022), and Zhang et al. (2022) in object classification accuracy across datasets of varying sizes. Its innovative features significantly boost object recognition and contextual analysis, resulting in more precise classifications.

17.5.1 ANALYSIS OF UNDISCOVERED RELATIONSHIPS

Although the suggested model exhibits improved classification efficiency, its capability to examine relationships that are not visible needs to be assessed in real-time scenarios. This efficiency was also gauged in relation to precision and accuracy, and a comparison was made with existing models in similar contexts. For example, Figure 17.4 illustrates the precision comparison of the model.

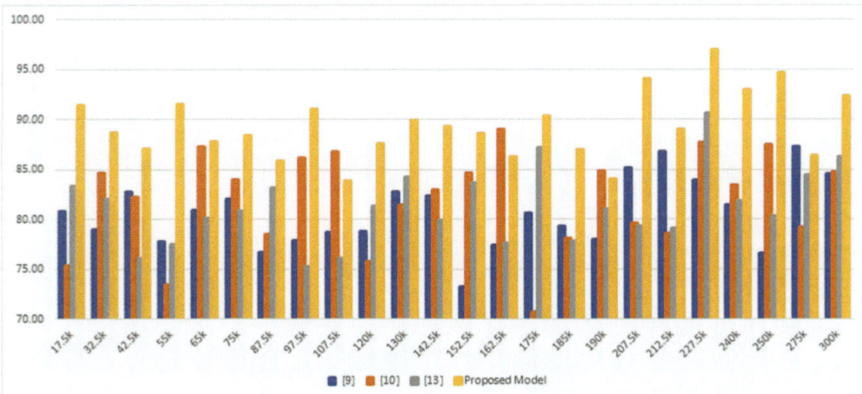

FIGURE 17.4 Precision in uncovering unseen relationships between objects.

The precision observed in analyzing novel relationships among detected objects is a critical metric that assesses a model's capacity to accurately recognize and articulate unfamiliar connections between objects in real-time settings. Let's examine the precision outcomes for relationship analysis across various dataset sizes among different models, including Zha et al. (2022), Huang et al. (2022), Zhang et al. (2022), and the proposed model.

The proposed model consistently achieves superior precision in analyzing unseen relationships compared to alternative models, regardless of dataset size. This capability reflects its effectiveness in precisely finding and describing novel connections among detected objects. For example, when evaluated on a dataset consisting of 17.5k samples, the model outperformed Zha et al. (2022) by 10.62%, Huang et al. (2022) by 16.02%, and Zhang et al. (2022) by 8.05%. This significant performance gap emphasizes the model's proficiency in recognizing and articulating unfamiliar relationships, even in situations where time is of the essence.

Additionally, the multimodal attention mechanism in the model improves its precision by facilitating the generation of clear and accurate captions that effectively communicate new relationships. In summary, the proposed model consistently excels in precision when detecting undiscovered relationships, demonstrating its capability to identify and articulate unfamiliar connections, further enhanced by the multimodal attention mechanism. In a similar manner, the effectiveness of the models is evaluated in Figure 17.5.

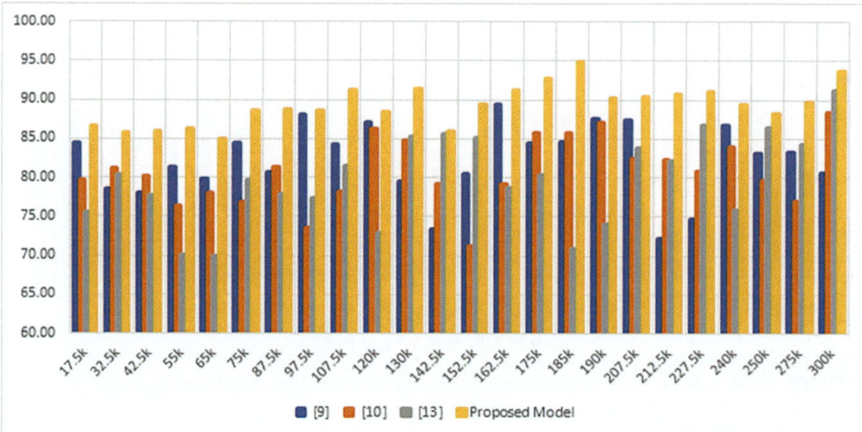

FIGURE 17.5 Accuracy in uncovering unseen relationships between objects.

The accuracy observed in discerning novel relationships among detected objects is a pivotal metric that gauges the model's capacity to accurately classify and articulate unfamiliar connections in real-time conditions. Let's examine the accuracy outcomes for relationship analysis among various models, including Zha et al. (2022), Huang et al. (2022), Zhang et al. (2022), and the newly proposed model, across datasets of varying sizes.

The proposed model consistently achieves higher accuracy in identifying unseen connections compared to other models across different dataset sizes. This demonstrates the system's effectiveness in accurately categorizing and representing new associations between identified elements.

For example, the proposed model's accuracy rate on a dataset of 17.5k samples is 86.71%, surpassing Zha et al. (2022) by 2.16%, Huang et al. (2022) by 11.99%, and Zhang et al. (2022) by 11.08%. This substantial performance gap suggests that the proposed model excels in accurately identifying and articulating unfamiliar relationships, even in real-time scenarios.

Furthermore, the proposed model's multimodal attention mechanism improves its precision by allowing it to produce captions that effectively and clearly articulate novel associations. In conclusion, the proposed model consistently surpasses other models in accuracy when detecting unseen relationships among objects, regardless of dataset size. This emphasizes its capability to effectively classify and convey unfamiliar connections, further strengthened by the multimodal attention mechanism.

17.5.2 QUALITATIVE RESULTS

The qualitative results are thoroughly analyzed. In Figure 17.6, panel (a) shows that both the relationships "going down" and "snowboarding down" are accurate predictions.

Input Image	**Generated Caption**
	A person in black snowboarding down a snowy hill.
	Two people running towards a soccer ball on a grass field.

FIGURE 17.6 Evaluation of the proposed model's qualitative performance.

However, "going down" is part of the training data, while "snowboarding down" is not. Similarly, in panel (b), the predictions "running after" and "running towards" are both plausible, though "running after" is present in the training data, whereas "running towards" is absent. Nevertheless, the proposed model effectively identifies and interprets these unseen relationships. This underscores the model's ability to produce precise and unique captions, as well as to connect them through relational analysis.

17.6 CONCLUSION

The proposed DL model was assessed for its ability to classify objects and analyze unseen relationships between detected objects within semantic text categories. Multiple performance metrics were used to gauge effectiveness, and the model's performance was compared against top models such as Zha et al. (2022), Huang et al. (2022), and Zhang et al. (2022). Comprehensive experiments were performed to thoroughly evaluate its capabilities. The model consistently demonstrated exceptional classification efficiency, outperforming existing models across datasets of varying sizes. Its precision and accuracy metrics highlighted its superiority, with accuracy reaching 91.25% on larger datasets. The model's proficiency in semantic text classification indicates its potential to revolutionize various fields, from computational linguistics to visual data analysis, by facilitating accurate and efficient object categorization.

17.7 FUTURE SCOPE

Although our study has yielded valuable insights into the capabilities of the proposed model for object recognition and relationship analysis, there are several promising avenues for future exploration that could expand the model's impact and versatility:

1. **Improved Semantic Comprehension:** Advancing the model's ability to grasp nuanced semantics.
2. **Dynamic Learning:** Implementing continuous learning methods to adapt to evolving scenarios.
3. **Real-Time Processing:** Enhancing the model's efficiency for instant applications.
4. **Collaboration between Humans and Machines:** Exploring ways in which the model can collaborate with human experts.

5. **Multilingual and Cross-Cultural Applications:** Assessing its adaptability to diverse languages and cultures.
6. **Hardware Optimization:** Utilizing specialized hardware for performance enhancement.

DECLARATION

No AI tools or any other similar tools have been used in the content creation of this chapter.

KEYWORDS

- **cross-cultural applications**
- **dynamic contextual graph networks**
- **dynamic learning**
- **hardware optimization**
- **image caption generation**
- **multimodal attention mechanism**
- **real-time processing**
- **unseen relationship detection**
- **zero-shot detection**

REFERENCES

Aafaq, N., Mian, A., Liu, W., Akhtar, N., & Shah, M. (2022). Cross-domain modality fusion for dense video captioning. *IEEE Transactions on Artificial Intelligence, 3*(5), 763–777. https://doi.org/10.1109/TAI.2021.3134190 (accessed on 11th July 2025).

Bae, J. W., Lee, S. H., Kim, W. Y., & Seong, J. H., & Seo, D. H. (2022). Image captioning model using part-of-speech guidance module for description with diverse vocabulary. *IEEE Access, 10*, 45219–45229. https://doi.org/10.1109/ACCESS.2022.3169781 (accessed on 11th July 2025).

Cheng, Q., Huang, H., Xu, Y., Zhou, Y., Li, H., & Wang, Z. (2022). NWPU-Captions dataset and MLCA-Net for remote sensing image captioning. *IEEE Transactions on Geoscience and Remote Sensing, 60*, 1–19, Art. no. 5629419. https://doi.org/10.1109/TGRS.2022.3201474 (accessed on 11th July 2025).

Chu, Y., Yue, X., Yu, L., Sergei, M., & Wang, Z. (2020). Automatic image captioning based on ResNet50 and LSTM with soft attention. *Wireless Communications and Mobile Computing*. https://doi.org/10.1155/2020/8909458 (accessed on 11th July 2025).

Hafeth, D. A., Kollias, S., & Ghafoor, M. (2023). Semantic representations with attention networks for boosting image captioning. *IEEE Access, 11*, 40230–40239. https://doi.org/10.1109/ACCESS.2023.3268744 (accessed on 11th July 2025).

Haque, A. U., Ghani, S., & Saeed, M. (2021). Image captioning with positional and geometrical semantics. *IEEE Access, 9*, 160917–160925. https://doi.org/10.1109/ACCESS.2021.3131343 (accessed on 11th July 2025).

Hartatik, H., Al Fatta, H., & Fajar, U. (2019). Captioning image using convolutional neural network (CNN) and long-short term memory (LSTM). In *International Seminar on Research of Information Technology and Intelligent Systems (ISRITI)* (pp. 263–268). https://doi.org/10.1109/ISRITI48646.2019.9034562 (accessed on 11th July 2025).

Hinz, T., Heinrich, S., & Wermter, S. (2022). Semantic object accuracy for generative text-to-image synthesis. *IEEE Transactions on Pattern Analysis and Machine Intelligence, 44*(3), 1552–1565. https://doi.org/10.1109/TPAMI.2020.3021209 (accessed on 11th July 2025).

Huang, Q., et al. (2022). Image difference captioning with instance-level fine-grained feature representation. *IEEE Transactions on Multimedia, 24*, 2004–2017. https://doi.org/10.1109/TMM.2021.3074803 (accessed on 11th July 2025).

Huang, W., & Wang, Q., & Li, X. (2021). Denoising-based multiscale feature fusion for remote sensing image captioning. *IEEE Geoscience and Remote Sensing Letters, 18*(3), 436–440. https://doi.org/10.1109/LGRS.2020.2980933 (accessed on 11th July 2025).

Im, S. K., & Chan, K. H. (2023). Context-adaptive-based image captioning by Bi-CARU. *IEEE Access, 11*, 84934–84943. https://doi.org/10.1109/ACCESS.2023.3302512 (accessed on 11th July 2025).

Kim, D. J., Oh, T. H., Choi, J., & Kweon, I. S. (2022). Dense relational image captioning via multi-task triple-stream networks. *IEEE Transactions on Pattern Analysis and Machine Intelligence, 44*(11), 7348–7362. https://doi.org/10.1109/TPAMI.2021.3119754 (accessed on 11th July 2025).

Liu, C., Zhao, R., & Shi, Z. (2022). Remote-sensing image captioning based on multilayer aggregated transformer. *IEEE Geoscience and Remote Sensing Letters, 19*, 1–5, Art. no. 6506605. https://doi.org/10.1109/LGRS.2022.3150957 (accessed on 11th July 2025).

Liu, C., Zhao, R., Chen, H., Zou, Z., & Shi, Z. (2022). Remote sensing image change captioning with dual-branch transformers: A new method and a large-scale dataset. *IEEE Transactions on Geoscience and Remote Sensing, 60*, 1–20, Art. no. 5633520. https://doi.org/10.1109/TGRS.2022.3218921 (accessed on 11th July 2025).

Liu, M., Hu, H., Li, L., Yu, Y., & Guan, W. (2022). Chinese image caption generation via visual attention and topic modeling. *IEEE Transactions on Cybernetics, 52*(2), 1247–1257. https://doi.org/10.1109/TCYB.2020.2997034 (accessed on 11th July 2025).

Ma, X., Zhao, R., & Shi, Z. (2021). Multiscale methods for optical remote-sensing image captioning. *IEEE Geoscience and Remote Sensing Letters, 18*(11), 2001–2005. https://doi.org/10.1109/LGRS.2020.3009243 (accessed on 11th July 2025).

Nguyen, K., Bui, D. C., Trinh, T., & Vo, N. D. (2022). EAES: Effective augmented embedding spaces for text-based image captioning. *IEEE Access, 10*, 32443–32452. https://doi.org/10.1109/ACCESS.2022.3158763 (accessed on 11th July 2025).

Nguyen, T. S., & Fernando, B. (2022). Effective multimodal encoding for image paragraph captioning. *IEEE Transactions on Image Processing, 31*, 6381–6395. https://doi.org/10.1109/TIP.2022.3211467 (accessed on 11th July 2025).

Sargar, O., & Kinger, S. (2021). Image captioning methods and metrics. In *International Conference on Emerging Smart Computing and Informatics (ESCI)* (pp. 522–526). https://doi.org/10.1109/ESCI50559.2021.9396839 (accessed on 11th July 2025).

Song, P., Guo, D., Zhou, J., Xu, M., & Wang, M. (2023). Memorial GAN with joint semantic optimization for unpaired image captioning. *IEEE Transactions on Cybernetics, 53*(7), 4388–4399. https://doi.org/10.1109/TCYB.2022.3175012 (accessed on 11th July 2025).

Sumbul, G., Nayak, S., & Demir, B. (2021). SD-RSIC: Summarization-driven deep remote sensing image captioning. *IEEE Transactions on Geoscience and Remote Sensing, 59*(8), 6922–6934. https://doi.org/10.1109/TGRS.2020.3031111 (accessed on 11th July 2025).

Thakare, Y. A., Walse, K. H., & Atique, M. (2024). AFSDCGN: Adaptive feature scaling and dynamic contextual graph networks for image captioning with unseen relationship detection. *Multimedia Tools and Applications*.

Wang, D., Hu, Z., Zhou, Y., Hong, R., & Wang, M. (2023). A text-guided generation and refinement model for image captioning. *IEEE Transactions on Multimedia, 25*, 2966–2977. https://doi.org/10.1109/TMM.2022.3154149 (accessed on 11th July 2025).

Wang, J., Xu, W., Wang, Q., & Chan, A. B. (2023). On distinctive image captioning via comparing and reweighting. *IEEE Transactions on Pattern Analysis and Machine Intelligence, 45*(2), 2088–2103. https://doi.org/10.1109/TPAMI.2022.3159811 (accessed on 11th July 2025).

Wang, Q., Wan, J., & Chan, A. B. (2022). On diversity in image captioning: Metrics and methods. *IEEE Transactions on Pattern Analysis and Machine Intelligence, 44*(2), 1035–1049. https://doi.org/10.1109/TPAMI.2020.3013834 (accessed on 11th July 2025).

Wang, Y., Zhang, W., Zhang, Z., Gao, X., & Sun, X. (2022). Multiscale multiinteraction network for remote sensing image captioning. *IEEE Journal of Selected Topics in Applied Earth Observations and Remote Sensing, 15*, 2154–2165. https://doi.org/10.1109/JSTARS.2022.3153636 (accessed on 11th July 2025).

Wu, J., Chen, T., Wu, H., Yang, Z., Luo, G., & Lin, L. (2021). Fine-grained image captioning with global-local discriminative objective. *IEEE Transactions on Multimedia, 23*, 2413–2427. https://doi.org/10.1109/TMM.2020.3011317 (accessed on 11th July 2025).

Yang, W., Wu, J., Tian, S., Li, L., Dong, W., & Shi, G. (2022). Fine-grained image quality caption with hierarchical semantics degradation. *IEEE Transactions on Image Processing, 31*, 3578–3590. https://doi.org/10.1109/TIP.2022.3171445 (accessed on 11th July 2025).

Ye, X., et al. (2022). A joint-training two-stage method for remote sensing image captioning. *IEEE Transactions on Geoscience and Remote Sensing, 60*, 1–16, Art. no. 4709616. https://doi.org/10.1109/TGRS.2022.3224244 (accessed on 11th July 2025).

Zha, Z. J., Liu, D., Zhang, H., Zhang, Y., & Wu, F. (2022). Context-aware visual policy network for fine-grained image captioning. *IEEE Transactions on Pattern Analysis and Machine Intelligence, 44*(2), 710–722. https://doi.org/10.1109/TPAMI.2019.2909864 (accessed on 11th July 2025).

Zhang, Y., et al. (2023). ACTUAL: Audio captioning with caption feature space regularization. *IEEE/ACM Transactions on Audio, Speech, and Language Processing,*

31, 2643–2657. https://doi.org/10.1109/TASLP.2023.3293015 (accessed on 11th July 2025).

Zhang, Z., Zhang, W., Yan, M., Gao, X., Fu, K., & Sun, X. (2022). Global visual feature and linguistic state guided attention for remote sensing image captioning. *IEEE Transactions on Geoscience and Remote Sensing, 60*, 1–16, Art. no. 5615216. https://doi.org/10.1109/TGRS.2021.3132095 (accessed on 11th July 2025).

Zhuang, S., Wang, P., Wang, G., Wang, D., Chen, J., & Gao, F. (2022). Improving remote sensing image captioning by combining grid features and transformer. *IEEE Geoscience and Remote Sensing Letters, 19*, 1–5, Art. no. 6504905. https://doi.org/10.1109/LGRS.2021.3135711 (accessed on 11th July 2025).

Index

C

G

M

Q

R

T

For Product Safety Concerns and Information please contact our
EU representative GPSR@taylorandfrancis.com Taylor & Francis
Verlag GmbH, Kaufingerstraße 24, 80331 München, Germany